SMALL BUILDINGS

SMALL BUILDINGS

Percy W. Blandford

TAB **TAB BOOKS**
Blue Ridge Summit, PA

FIRST EDITION
THIRD PRINTING

Library of Congress Cataloging in Publication Data

Blandford, Percy W.
    Small buildings.

    1. Outbuildings—Design and construction—Amateurs'
manuals.   I. Title.
TH4955.B57 1989   690'.89   88-35945
ISBN 0-8306-9144-8
ISBN 0-8306-3144-5 (pbk.)

TAB BOOKS offers software for sale. For information and a catalog, please contact
TAB Software Department, Blue Ridge Summit, PA 17294-0850.

Questions regarding the content of this book should be addressed to:

    Reader Inquiry Branch
    TAB BOOKS
    Blue Ridge Summit, PA 17294-0214

Edited by Nina E. Barr
Cover photograph courtesy of Jer Manufacturing Inc.

CONTENTS

INTRODUCTION

Have you ever thought that you couldn't construct a building? Building a home is very complicated, with all the plumbing, drains, and electrical work. A small building, such as a storage shed, a shop, a studio, a bird house or other animal house or run, or a stable, however, are all possible without much skill or advanced equipment. This book demonstrates how you can do all the construction work.

If you are an amateur woodworker, most of your projects probably have been smaller items of furniture, toys for the children, built-in items for the home—all of which you can handle on a bench or at least within the confines of a room or shop. A building is obviously too large to be dealt with in that way and that might make the prospect seem rather daunting. You do not need to feel this way. You certainly need space if you are to prefabricate parts before erecting on-site. You can do much of that work alongside your shop. It is certainly helpful to have your hand and power tools within reach, but if you remember to take all you need, you can do all the work in the place where you want to locate the building.

The size of your project need not be a problem. It is just as easy to cut joints on a piece of wood 15 feet long as it is on a piece 2 feet long. In fact, much of the construction of a building is simpler than making many furniture items. Do not hesitate to make any small building simply because it is bigger than anything you have ever made. I hope to show you in the following pages that you are capable of making a building of any reasonable size.

You might be worried that you will not finish with a square, symmetrical building. It is a fact that your approach to marking and setting out must be different. You can use squares and rules on individual pieces of wood, but they are of little use when you need to shape a 20-foot square. Actually, the techniques for dealing with shapes of these larger sizes are simple and interesting, as you will discover in later pages. Something like the hexagonal gazebo might be a little more complicated, but even that is simple if you follow the steps given.

You might think you do not have all the tools and equipment to begin building. Have you made simple furniture? Can you make boxes, toys, household items, and other small wooden assemblies? If so, you have all the skill and most of the equipment needed. You can make other things you need—and they are not many.

Most of the actual constructional work, even if it is large, is simpler than many bench jobs, so you will need fewer tools. You will certainly make a lot of use of a hammer! If you do not have a portable saw and electricity within reach, none of the wood is likely to be so large that handsawing is arduous. An electric drill is useful, and you probably have one. Many small buildings have been made with just the tools that you can carry in a small box. If you read through the instructions for the building of your choice and think what tools you need for each stage, you almost certainly will discover that you have them already.

Making and erecting small buildings is a branch of woodworking different from most others, mainly because the work is less compact. Do not let that fact put you off. You need a site, and you might need to level it and pour concrete. You need to plan ahead the steps in assembly and erection. You might need help handling some of the subassemblies. All of these processes involve the woodwork you know already, but on a larger scale.

If you make one or more small buildings for your own use and this sort of woodworking appeals to you, neighbors might ask you to make buildings for them. If you wish, you could find yourself fully occupied in a custom-building small business. With individual work, you can fit a building to suit needs or a particular space or situation. You can place doors and windows where they fit best. This building will be much better than a mass-produced sectional building that must be a set design.

Making a small wooden building is a very satisfying form of carpentry. At the end of the work, your product is certainly large enough for others to see. You can look at it and say, "I made that." Others will look at it and ask where you got such a marvelous building. Your reply will boost your ego.

Of my many woodworking activities, I have found making small buildings among the most satisfying. I built my own workshop many years ago. It still encourages me to tackle work in it to a high standard, knowing that it is not only well built, but that I planned its size, shape, and arrangement of windows and doors to make the best possible use of the situation. No factory-produced sectional building could have competed.

I hope you will find buildings described in the following pages that will show you the work is within your scope. In a short time, you will have the satisfaction of looking at and using your own shed, shop, playhouse, barn, or whatever appeals to you.

Note: Unless otherwise indicated, sizes on drawings and in material lists are in inches. Widths and thicknesses quoted are nominal, but lengths are mostly a little full.

1
PREPARATIONS

EVEN THE SMALLEST BUILDING IS USUALLY BIGGER THAN THE MAJORITY OF WOOD-working projects you normally make, and its size introduces a few special considerations. Instead of doing layout and squaring assemblies on the bench, you are faced with floor areas and structures larger than your normal equipment can span. It would be unwise, for instance, to use a 12-inch square and extend its line to 6 feet, as the possible error at the limit could be more than would be acceptable. Instead, it is better to use geometric methods, preferably to a size larger than the final result has to be so you can avoid possible errors.

It should be safe to assume that the corners of a sheet of 4- × -8-foot plywood are square, so within the limits of that size, you can use a sheet of plywood for marking and checking corners. It might be worthwhile making a 45-degree triangle by cutting from the corner of a sheet equal lengths along each side. You could use the edge of a sheet of plywood as a straightedge, but it would be better to find your straightest piece of wood of greater length than that and use it as a straightedge. You can check its straightness by drawing a line against it on a flat surface, then turning it over to see if the line matches (FIG. 1-1A).

If you want to mark long lines, it is better to use a chalk line. Have a piece of fine line (crochet cotton is suitable) and rub it with chalk, without jerking it so the chalk is shaken off. Have an assistant hold one end down, or use an awl (FIG. 1-1B). Stretch the line and hold it down. Reach as far along as you can, then lift the line a few inches and let it go, to deposit a fine line of chalk on the floor (FIG. 1-1C). If the length is so great that you cannot reach near the middle of the line, get someone else to "snap" the line at its center.

If you need to get a large corner square, as you would for a foundation or floor, draw a baseline longer than you will need. On it mark where the square line is to come (FIG. 1-1D). Now use the geometric property of right-angled (90 degrees) triangles with sides 3:4:5 proportionally. The right angle is between the two shorter sides.

Choose sizes for the sides of the triangle that will result in the square line you draw being at least as long as the final size you need. Suppose you need

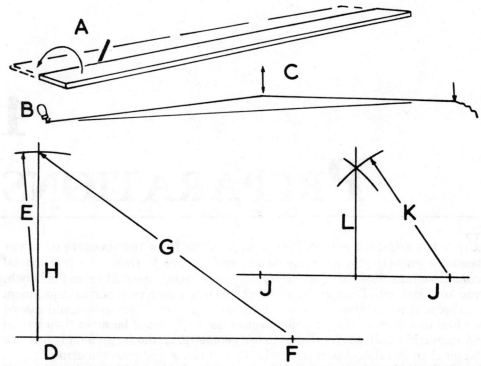

FIG. 1-1. *Check straightness by turning a board over (A). Striking a cord (B) will produce a long line. Make right angles by measuring (D,E,F,G,H) or drawing arcs (J,K,L).*

12 feet. If that is the "3" side of the triangle, the unit to use is 4 feet, making the triangle sides 12 feet, 16 feet, and 20 feet. Use a steel tape measure or other convenient means of measuring to draw a short arc from the point on the baseline at 12 feet radius (FIG. 1-1E). The arc should swing over what obviously will be the position of the square line. From the point, measure 16 feet along the baseline (FIG. 1-1F). From that point, measure 20 feet to a position on the arc (FIG. 1-1G). From the starting point, draw a line through the mark on the arc with a chalk line or a long straightedge (FIG. 1-1H). This line will be square to the baseline. Measure other lines parallel to it or the baseline.

If you want to erect a line square to the baseline away from a corner, you can measure equal distances on each side of where it is to be (FIG. 1-1J), then swing arcs from these points (FIG. 1-1K). Draw your square line through the point on the baseline and the crossing of the arcs (FIG. 1-1L). Arrange sizes so the radii of the arcs come at about 45 degrees to the baseline to get a crossing high enough to give a length of square line as big as you need.

The best way to check the squareness of anything you assemble, when opposite sides are the same length, is to compare diagonal measurements. In an ordinary rectangular frame, measure corner to corner (FIG. 1-2A) and adjust the frame until these lengths are the same. You do not have to use the extreme corners. If it is more convenient, take other points that should be square or

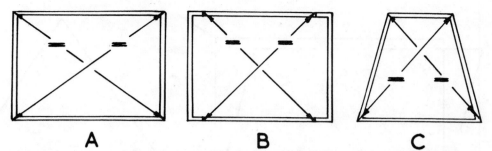

FIG. 1-2. To check squareness and symmetry, compare diagonal measurements.

measure along the sides from the corners (FIG. 1-2B). You can use the same technique to check the symmetry of anything that is not square (FIG. 1-2C). As assembly of a building progresses, you can compare diagonal or other measurements that should be the same. As the building takes a cubic form, you can compare the distance from the top corner of one side with the bottom corner of another.

RIGIDITY OF SMALL BUILDINGS

It is useless trying to achieve squareness if what you make is so poorly designed or made that it goes out of shape. The ability to hold a shape depends mainly on *triangulation*. If you join four pieces loosely at the corners and one corner is pushed, the frame will go out of shape (FIG. 1-3A). If you make a three-piece frame (FIG. 1-3B), nothing can push it out of shape. Put a diagonal piece across the four-sided frame and you have two triangles (FIG. 1-3C). Providing no parts bend, this framework will hold its shape. Smaller triangles might be sufficient (FIG. 1-3D). If you cover a framework with plywood, you have thoroughly triangulated it, and its shape will hold, but if you nail boards across, movement is still possible and it is advisable to add one or more struts to provide triangulation.

In a roof truss, you have the rigidity of a triangle, whether the tie in is at the eave's level or higher (FIG. 1-3E) to provide headroom. Any other framing in a roof truss is there to provide stiffness, but it also contributes to rigidity.

In a large roof, you support the covering on lengthwise purlins attached to the rafters. The assembly is comparable to a squared framework. Some stiffness comes from the building below, but triangulating the roof will relieve that of the excessive load. If the roof covering is made up of large sheets of stiff plywood rigidly fixed down, that will be sufficient, but if the covering is made of many sheets of corrugated metal or plastic, or something similar, movement is possible and diagonal *wind bracing* (FIG. 1-3F) might be advisable.

Similar needs for triangulation might be necessary in smaller assemblies, such as doors. If you make a door with several upright boards and ledges nailed across (Fig. 1-3G), the door will soon sag. To prevent this sagging, add braces (Fig. 1-3H), preferably notched into the ledges. If the door tries to sag, the compression loads on the braces stop it.

3

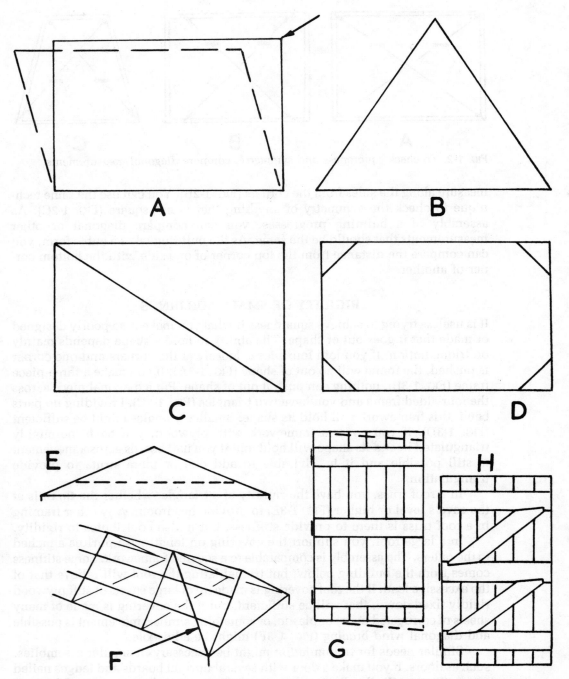

FIG. 1-3. A four-sided assembly might push out of shape (A). A three-sided figure will not push out of shape (B). Stiffen four-sided frames by triangulating them (C,D,E,F). To prevent a door from sagging, use diagonal braces (G,H).

If you are making the side of a shed or something similar, consider all the parts that are being built in, such as windows and doors, or anything that will be inside and attached. If there will not be an absolutely rigid skin and the lines are all parallel in two directions, add some diagonal struts to keep the assembly in shape. Long struts are most effective, but smaller, diagonal braces can give considerable stiffness.

FOUNDATIONS OF SMALL BUILDINGS

Rarely do you erect a building directly on the ground. Usually you have to prepare a base on which it stands and to support it. You might want to use a sectional building temporarily, such as a shed for garden tools during the summer months only, or you will use the building only briefly and then move it to another position. A dirt floor then might be acceptable, but you still must level it. Even a temporary building that is obviously not level looks wrong to you and all other observers. If you erect walls out of true, you will have difficulty fitting the roof.

Compacted soil might take the weight of a small building, but you must ram it or roll it hard before you put the building on it. If soil settles under the weight of a wall, you might have difficulty lifting and supporting it to make it level.

Even if you are satisfied with a dirt floor, it is usually better to arrange more solid supports under the walls. You could dig out a shallow trench and fill it with sand and small stones, rammed tight (FIG. 1-4A). It would be better to use concrete for the top few inches (FIG. 1-4B). You can embed anchor bolts as shown in FIG. 1-4C. With stones only, you will have to drive spikes through them into the ground.

It is possible to use bricks or concrete blocks. It might be satisfactory to put them in line (FIG. 1-4D), but you can spread the load better if you place them crosswise (FIG. 1-4E). Wood as a foundation material might rot. Some woods, however, have a good resistance to rot. You can expect other wood, which is soaked in preservative, to have a reasonable life. However, painting preservative on your wood does not achieve much penetration and would have little effect. You can use old railroad ties, since they are saturated with preservative and should be immune to rot.

It is more usual to use an all-over concrete pad as a foundation for a wooden building. You can extend it outside to form a path or patio. You also can use it as the floor of the building if that will suit its use. A floor made of wood or other materials is also a good seal against moisture and rodents. The thickness and the way you form the foundation will depend on the state of the ground. On a hard soil, you could cover a few inches of rammed, small stones with 3 inches of concrete, but if the soil is loose and sandy, you would need to increase both thicknesses.

Most ground is not as level as you would wish. If you want to put the building on the side of a hill, the slope is obvious. You possibly will have to make a foundation pad partly into the slope at the back and build it up at the front.

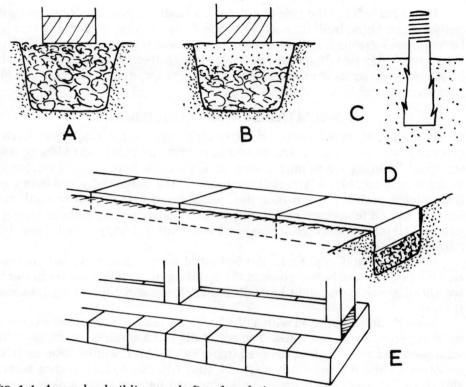

FIG. 1-4. *A wooden building needs firm foundations.*

On apparently level ground you might find a few inches difference in the length of the building, and you must allow for that. You must consider the circumstances and decide if you want one end above the surrounding ground or if you should go deeper at the other end. A compromise might be more appropriate. This decision applies whether you are putting down foundations under the walls only or laying an over-all pad.

For levelling, use the longest possible level, but supplement it with a long, straight board, preferably to span the whole foundation diagonally. It would be unwise to use the level alone to check a distance much greater than its own length.

Start with one side. If you will be putting down blocks, drive in several pegs as guides so their tops are level (FIG. 1-5A). You can use this procedure for a concrete foundation, or you might put down the shuttering strip, making sure its top is level with what will be the top surface of the concrete (FIG. 1-5B). Use pegs on both sides of the strip and pack it so it cannot move once you have it level and straight.

Next, work square to the first side, setting out the angle as described earlier. Level the pegs or shuttering in the same way. As a final check on that side, put your levelling strip across diagonally (FIG. 1-5C). If your levelling strip does not

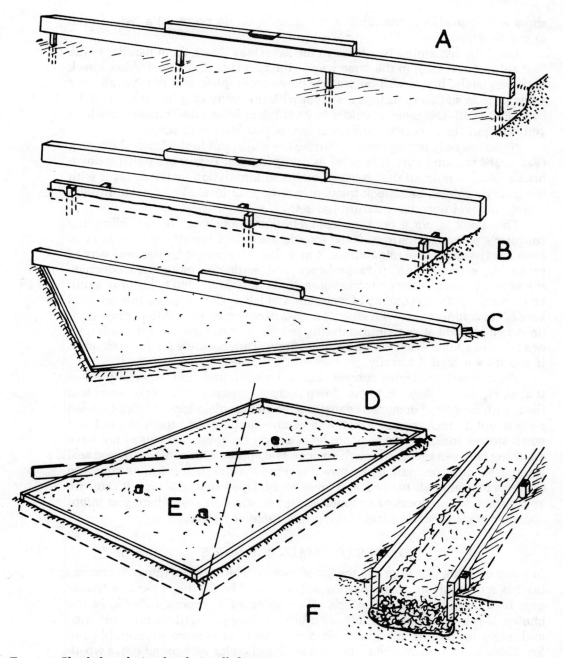

FIG. 1-5. Check foundation levels in all directions.

show a true level, test the other ways again—you do not want a twist in the foundation.

Level the remaining two sides in the same way. Mark them parallel to the first sides. Level each in the same way as the first side (FIG. 1-5D). Also, check both diagonals. If you will be putting down a complete pad of concrete, put a few pegs in so that their tops are level with the shuttering in the body of the base (FIG. 1-5E). Use them as guides to levelling or laying the stones over which you will put the concrete. Pull them out as you lay the concrete.

If you are only laying concrete under the walls, put in the inner shuttering (FIG. 1-5F), making sure it is level as you progress. If you are putting down bricks, blocks, railroad ties, or other sectional-foundation material, lay it with the pegs as guides, but check frequently with your level. You can do little to correct unlevel once the mixture has set.

This book is not a book on concrete work, and you should follow the supplier's recommendations. If the foundation is in a position where you can have ready-mixed concrete delivered and shot, that might be the best way of preparing a foundation. You probably will settle for more than adequate thickness. If you mix the concrete yourself, do not be tempted to lay only a thin layer, for the sake of economy of materials and labor. Have a good, consolidated base of sand and stones, with about 3 inches of concrete, even if the only weight on it will be light storage and standing people. For storage of yard machinery or a car, use an increased thickness of concrete. Thin concrete might crack, even if you do not load it heavily.

An alternative to laying concrete as a one-piece foundation, especially where it also will be the floor, is to put down precast concrete slabs. You can obtain them with decorated or stonelike surfaces, as well as plain tops. You can spread slabs about 24 inches square over a large area quickly. Bed them in sand and small stones, making sure they are level as you progress. They need not have anything between them, although you can seal spaces with concrete. If you seal the spaces with concrete, make sure it is more than ½ inch thick, as very thin concrete mortar tends to crack and break away. Open spaces filled with soil are appropriate to the floors of summer houses or sun lounges, where you might consider grass or small plants taking root there attractive.

FLOORS OF SMALL BUILDINGS

In many buildings the floor and the foundation will be the same thing. A concrete floor is acceptable for many purposes. If you will be driving a car or a tractor over it or there is a risk of spilling oil or water on it, concrete should be the choice. If it is a workshop or other building where you will spend some time and might drop tools or equipment, a wooden floor is more comfortable and less liable to damage dropped tools. If it is a building for year-round use, a wooden floor gives a more equitable temperature underfoot. You can coat concrete with rubberized or plastic sealants that might give you ¼ inch of insulation and a more comfortable surface, but these are not as satisfactory as wood for long-term standing.

The usual wooden floor in a small building might be very similar to the floors in many houses, with boards laid over joists (FIG. 1-6A). Sections of wood and spacing of the wood will depend on the size of the floor and the amount of support needed. As a guide, you can support boards which are ⅞ inch thick on 2- × -4-inch joists at 15-inch centers in an average small building (FIG. 1-6B). In a very small building, the joists need only be 3 inches deep. Particleboard, at least ¾ inch thick, makes a good alternative to parallel boards. With joints on joists, you have the minimum of gaps (FIG. 1-6C).

Avoid joists or other wood resting directly on concrete, brick, or stone. If you are supporting the ends of joists at wall foundations, insulate them from moisture (FIG. 1-6D). This insulation might be material sold for damp-proofing, roof covering, or just pieces of plastic sheeting. If you intend to support the length of the joists by the concrete foundation, put strips of plastic under them or cover the whole concrete surface with plastic sheeting. It might be better to raise

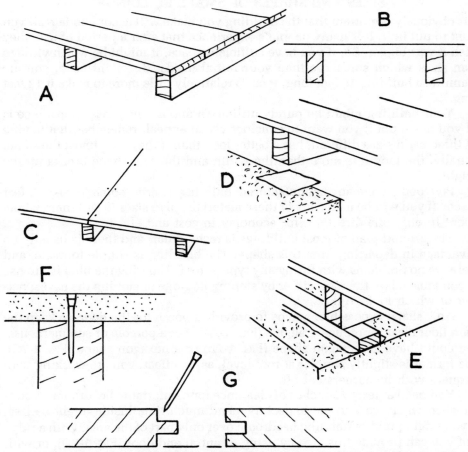

FIG. 1-6. A floor may be boards or particleboard on joists (A,B,C). Insulate wood from concrete (D,E). Conceal nails (F,G).

the wooden floor above the foundations especially if the concrete is not level in the main area. Even if the wooden floor starts apparently level, eventually it might take the uneven shape of the concrete below. A strip over damp-proofing might support the joists (FIG. 1-6E) above the concrete. In some buildings, the bottom member of the wall might support the joists.

Punch nails through particleboard below the surface (FIG. 1-6F). If you use plain boards, bring them tightly together and sink their nails below the surface. Better floor boards have tongue-and-groove joints. They will maintain a more even surface, which is important if you want to lay carpeting or other floor covering. You can conceal most of their nailing by driving diagonally through the tongues (FIG. 1-6G).

One surprisingly hard-wearing floor covering is hardboard. In damp conditions, use the oil-tempered variety. Fasten it down with plenty of fine pins or nails. An initial wax spray or polish should seal it for life.

SIZES AND SHAPES OF SMALL BUILDINGS

It is obviously important that the building you make will accommodate all you want to put in it. It is many people's experience that after a period of use, they wish it was bigger. So, if you have sufficient space, it might be worthwhile to plan sizes which suit more than your initial needs. If, for instance, you are planning a building 10 feet long, it costs relatively little more to make it 12 feet long.

Your building might be purely utilitarian and a functional appearance is all you need, but if you want to consider visual appeal, remember that having all three main sizes different has a better look than if they are almost the same. Usually, the length is more than the width and the roof shape breaks up the height.

Plywood and many manufactured building boards come in 4- × -8-feet sheets. If you will be using any of these materials, plan sizes to suit them whole or cut in only one direction, for economy in cost and effort.

The ground plan of most buildings is rectangular, and there is usually no advantage in departing from this shape. The building is simple to make, and there are no problems with fitting any type of roof. Tapering the plan is unwise as you must adapt the roof, either by sloping its ridge or sloping the eaves; neither of which looks right.

An L shape is possible, maybe to provide a porch for the door, to keep the main floor clear, or reduce draughts (FIG. 1-7A). For a pergola or summerhouse, you could have an octagon shape (FIG. 1-7B) or a hexagon shape (FIG. 1-7C). The former is slightly simpler at roof level, as, in effect, you are working with a square with its corners cut off.

You need at least 75 inches of clearance for comfortable headroom. A storage shed where you do not expect to spend much time inside might be less, or you might provide standing headroom over only part of the area. With a ridge roof you can provide this area by using a central end door (FIG. 1-7D), or with a sloping roof, you might have the door at the higher side of the building (FIG.

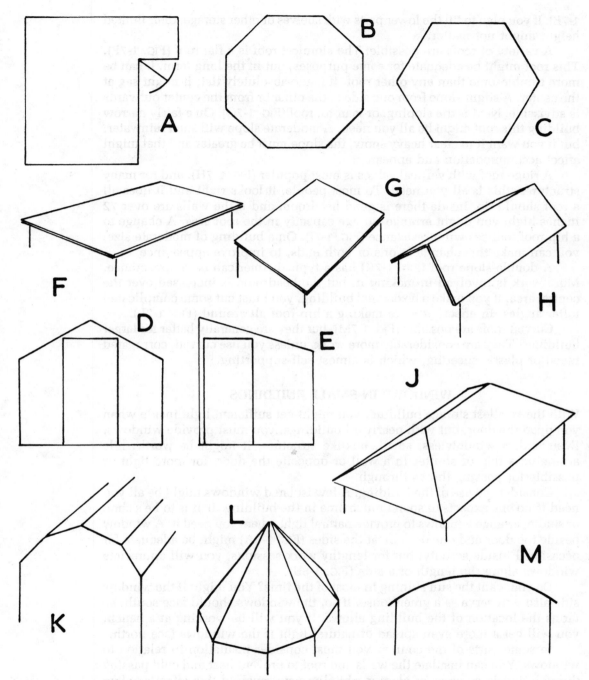

FIG. 1-7. You can plan buildings in many shapes and with different roofs.

11

1-7E). If you plan to fill the lower parts with shelves or other storage, their limited height might not matter.

A variety of roofs are possible. The simplest roof is a flat roof (FIG. 1-7F). This roof might be adequate for some purposes, but in the long term, it can be more troublesome than any other roof. If made absolutely flat, it might sag at the center. A slight slope from one side to the other or from the center outwards is advisable. Next is the sloping, or lean-to, roof (FIG. 1-7G). On a fairly narrow building this roof might be all you need. A moderate slope will shed rainwater, but if you want it to clear heavy snow, the slope must be greater and that might affect accommodation and appearance.

A ridge roof with vertical gables is most popular (FIG. 1-7H), and for many structures, this is all you need. To most people, it looks right and it does all a roof should do. Inside there is good headroom and if the walls are over 72 inches high, you might arrange storage capacity in the roof area. A change to a hip roof reduces wind resistance (FIG. 1-7J). On a building of moderate size, you can make this change at one or both ends, to improve appearance.

A double-slope roof (FIG. 1-7K) has a typical American barn appearance. More work is involved in making it, but the headroom is increased over the central area. If you make a hexagonal building, you must cut some complicated rafter angles. In effect, you are making a hip roof all around (FIG. 1-7L).

Curved roofs are possible (FIG. 1-7M), but they are generally better for larger buildings. They are considerably more work, unless you use curved, corrugated metal or plastic sheeting, which is almost self-supporting.

WINDOWS IN SMALL BUILDINGS

With the smallest storage building, you might get sufficient light inside when you open the door, but with nearly all buildings, you must provide windows. Even with a windowless, small, boxlike structure, it might be worthwhile arranging a flap or shutter in a wall or opposite the door, for more light or possibly for passing things through.

Consider the use of the building. A few isolated windows might be all you need if no one expects to spend much time in the building. If it is to be a shop or studio, arrange windows to provide natural light where you need it. A window beside the door and one or more at the sides (FIG. 1-8A) might be adequate for occasional inside activity, but for lengthy work sessions, you will appreciate windows along the length of a side (FIG. 1-8B).

Do you want the sun shining in most of the time? You might if the window side also will serve as a greenhouse. If so, the windows should face south, so far as the location of the building allows. If you will be working at a bench, you will get a more even spread of natural light if the windows face north.

In some parts of the country you must consider insulation in relation to windows. You can insulate the walls and roof to prevent heat and cold passing through, but large areas of glass might eliminate much of this effect as glass has really no insulating properties. Double glazing would help, but for most small buildings, that is a complication you will want to avoid. It is better to

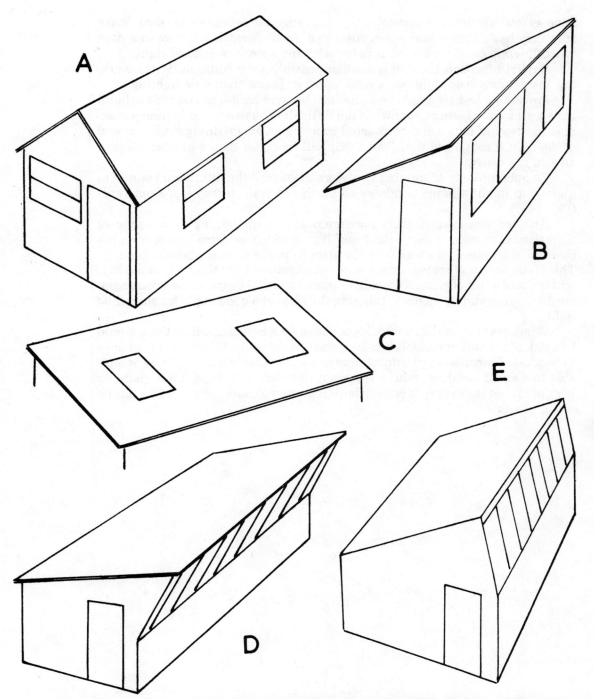

FIG. 1-8. *Arrange windows in walls (A,B) or roof (C). Sloping windows give different types of illumination (D,E).*

have as few windows as possible. If you arrange the windows to open, make sure they have a good seal when you close them. A small window in a door is worthwhile considering, as it helps with the spread of natural light.

Daylight through the roof is another possibility for sunlight. When working inside, overhead lighting is often more welcome than side lighting.

Preventing leakage might be a problem, but roof lights are available to build in as a package. Making your own is not difficult. You easily can fit transparent-plastic sheeting into metal-corrugated sheeting. A surprisingly small amount of daylight through the roof (FIG. 1-8C) will be better than a greater window area in the walls.

To obtain plenty of lighting in a work area, with the minimum amount of glare, you might arrange windows all along one wall so they slope outwards (FIG. 1-8D).

Another way of providing maximum and even lighting in a studio or workshop is to make a north-light roof (FIG. 1-8E). The name comes from the practice of arranging the roof with the glass facing as near as possible to north. This arrangement gives you an even spread of natural light throughout the day, without much risk of glare. If the ridge surfaces are at 90 degrees, the eaves angles are 30 degrees and 60 degrees, bringing the glass at a good angle for spreading light.

Windows only in the roof are a good idea for a building where there would be a risk of animals or vandals breaking glass in the walls. Windows do not have to be glass. Transparent plastic alternatives are available that might be acceptable in a small building, where you want light let in, but the finest visibility in or out is not important. If you are building a greenhouse, you will find plastic easier to use.

2
BUILDING
DETAILS

Y OU CAN USE SEVERAL MATERIALS TO MAKE SMALL BUILDINGS. BRICK OR STONE PRO-
vide the most substantial structures. You can make them completely of metal,
or they might have wooden frames and metal covering. Several plastic materials
are suitable for cladding and roofing over wood or metal supports. You can use
many manufactured boards with a wooden base in parts of a building. However, most small buildings are made of wood throughout, or with only small
amounts of other materials.

This book is for the woodworker who wants to make one or more small
buildings and prefers to use the material he knows. Most of the work is quite
straightforward and does not have the complications associated with making
good furniture or similar, better-quality woodworking. Obviously, skill is worth
having, but you can do much of the work successfully with a little skill and
by working carefully. The tools you need are not many. If you have a
comprehensive tool kit, you will find uses for it, but it is possible to make many
buildings with very few tools, particularly if you get all the wood sawed and
planed to size from a lumberyard.

You can use hand or power tools. It would not be very laborious to make
many small buildings entirely with the aid of hand tools. The advantage of power
is mainly in speeding up some operations. You might gain accuracy with some
power tools, particularly when you prepare parts in the workshop. On the site,
you are more likely to reach for a handsaw or plane, and you will make much
use of a hammer, for which there is no satisfactory power alternative. The most
useful power tool is an electric drill. A router with a few plain cutters will speed
up the making of joints. If you have a portable power saw, you will use it, but
you can cut most wood for your building easily with a handsaw.

MATERIALS

You can use almost any wood for a small building. Much will depend on what
is available in your area. Some woods are more durable than others, but
preservatives on the less-durable woods will give them a long life. In general,

there is no need to use hardwood. Most hardwoods are harder to work with than softwoods, and they are unnecessarily heavy. Most are considerably more expensive than softwoods. Some very resinous woods have a good resistance to rot and you can use a few without treatment on the outside. Consequently, they weather to a pleasant color that might blend into the surroundings. You should paint or treat most small, wooden buildings, however, so they are protected from the effects of sun and rain.

You can use any of the common softwoods. Some you will be offered might have a larger number of knots. A few small knots might be acceptable, but for the structural parts, try to get wood with reasonably straight grain and few knots. If a knot is black around its outside, it is loose and will not contribute to strength, even if it does not fall out. A knot without a black rim is less trouble. A knot falling out of cladding or siding would be a nuisance.

For many parts of most small buildings, you can use sawed wood, but where it will show, it is better to use planed wood. Planing at a mill will reduce sizes by about ¼ inch, and you must allow for that. For instance, 2- × -4-inch wood planed all around actually will be about 1¾ × 3¾ inches. If the structure will be hidden between the outside covering and an inner lining, you could use sawed wood and get the benefit of extra strength from the thicker wood. It also should be cheaper. Let your supplier see the material's list. If he can provide short pieces, you might get a better price and service than if you tried to buy long lengths to cut yourself.

Plywood

You can use plywood in many thicknesses for several parts of a small building. Sheets are mostly 4 × 8 feet, so it is wise to scheme building parts to use sheets whole or to cut economically. Any wood which can have the log rotated and peeled into thin veneers can be made into plywood. Much of the available plywood is made of Douglas fir, which is satisfactory for buildings. You can use many other woods also, some of which have a better appearance and take paint or other finish better.

Many grades of plywood are available. Plywood grades are determined mostly by appearance. Prices can vary considerably and there is no point in paying for plywood free from knots in both surface veneers if one side will be hidden.

More important is the purpose for which you intend to use the plywood. Today, most glues used in plywood have a good resistance to moisture, but this is not always true. General-purpose plywood might only be suitable for indoor use. However, the inside of a small building might also get wet, so it is better to avoid this plywood. External plywood has a waterproof glue. Marine-grade plywood is even better quality than external plywood, as the plies and their choice and arrangement, as well as the glue, suit boatbuilding. There is no need to pay the extra price for the marine plywood. For small buildings, exterior plywood is what you need.

Particleboard

As its name implies, particleboard is made of particles or chips of wood embedded in a resin. Boards are the same size as plywood. For our purpose, it should not be less than ¾ inch thick, for strength and stiffness. You can saw it and plane it, but it is unsuitable for cutting joints.

Most particleboard will suffer if exposed to moisture, so it is unsuitable for outside use, even when you paint it.

Particleboard will make good floors. With its large, overall coverage, it might be better in your building than a floor made of many comparatively narrow boards. Similarly, it makes good one-piece shelves, but would not be suitable in a greenhouse or other place where it would be wet for long periods.

You can nail or screw particleboard. For nails near an edge, it is advisable to drill undersized holes, to avoid breaking out. For screwing into particleboard, a tapping-size hole should be taken as far as the thread will go. A screw will not cut its own way to the full depth, as it will in wood.

Hardboard

Hardboard is made from compressed wood fibers and the sheets commonly available have one very smooth side and a patterned, opposite side. Sheets are the same size as plywood, but the thickness is only ⅛ inch.

The quality of hardboard available varies tremendously, and is largely dependent on its density. Some of it is little better than cardboard and will disintegrate in a similar way if you allow it to become wet. This hardboard, and even the better-quality, general-purpose hardboard, is unsuitable for use in a small building, except possibly for backs of cupboards or bottoms of drawers in the furnishings of the building. You could use it as a lining, when there is insulation material between it and the outer covering, but the oil-tempered type would be better.

Hardboard might be treated so it has an oil impregnation. This oil gives it a resistance to water. It is not waterproof if subjected to moisture for a long period. Trade names vary, but there is something in the name to indicate "oil-tempered" or a similar term. You can use oil-tempered hardboard for the outside surface of a building, but you must paint it constantly. Better coverings are available.

You can obtain hardboard sheets already perforated with a pattern of small holes. It might be called "peg-board" or something similar. The material is not usually oil-tempered. Many metal clips are available to hook into the holes. These clips will make tool racks. Peg board also will make a ventilated lining, but hardboard has only limited use in a better small building.

Insulation

For many purposes, it might not be necessary to consider insulation for a small building. If it is just a store for garden tools, the inside temperature might not be important. If it is a workshop for year-round use, you will want to keep a comfortable temperature whatever the condition is outside.

Wood in itself provides some insulation, better than metal or solid plastic. If you have wood outside and wood or hardboard lining, the air between also will provide a temperature barrier. You will improve this temperature barrier by adding one of the insulation materials used in home buildings, such as fiberglass. Wall insulation will be helpful, but roof insulation is important and not so easy to provide. Heat rises and much could escape through an uninsulated roof. A lining could hold insulation material directly under the roof, or you might fit a ceiling with insulation material above it. A wooden floor, providing the wind cannot blow under it, might be its own sufficient insulation.

Precautions to keep heat in also will work in reverse, if keeping cool is your problem. You can then have plenty of open windows and doors to increase ventilation. Whether heat or cold is your problem, ventilation is important, otherwise your building could suffer from condensation. Arrange ventilators low and high, so air can circulate. You can create a pattern of holes with flaps to cover, if the breeze is in the wrong direction, or you want to keep out rodents.

Wooden Sections

Most of the wood you use in a small building will be plain, rectangular sections and in stock sizes. For covering, boards are made that provide weatherproofing, even if they expand and contract. If the boards are to shed

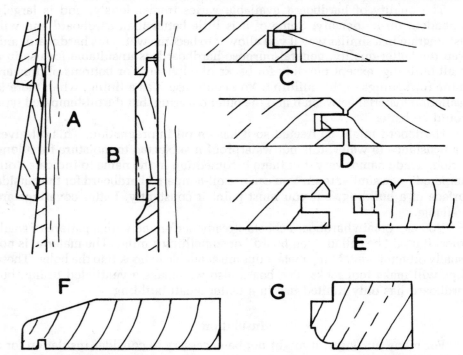

FIG. 2-1. Arrange or join boards in several ways to provide weatherproofness (A,B,C, D,E) A groove under a sill stops water from running underneath (F). You can mold a section with a rabbet (G).

water, arrange them horizontally so they overlap. One type has a tapered section (FIG. 2-1A). A better type fits flat and is cut to provide the overlap (FIG. 2-1B). This procedure might be called *shiplap*, from the way you lay planks of lapstrake boat skins.

If you are to lay the boards vertically, they might have tongue-and-groove joints. The plain joint is not as attractive as some others. If the wood shrinks, the gaps are very obvious (FIG. 2-1C). You can make this gap less obvious with chamfers (FIG. 2-1D). Another form which disguises the gap has a bead (FIG. 2-1E). Doors are usually made with vertical boards, even if the wall cladding is horizontal.

The window surround must protect wall-covering boards. In particular, rain must not run behind the boards. At the bottom, there should be a sill with its top sloping to shed water and a groove underneath to prevent water from running back (FIG. 2-1F). Stock sections are available.

Window glass is fitted best into a rabbet. If you do not prepare your own, you can buy a stock section. This section might be molded also (FIG. 2-1G). A rather similar section might be available for a door surround and another for a door step. Unfortunately, some stock sections are meant for home building and might be too big for a small, wooden building. Check what is available locally and you might be able to plan details of your building to use these standard pieces.

NAILING

You can nail many parts of a small building. There might be no need to do anything more elaborate than put one piece of wood on top of another and nail it there. Occasionally, screws might be more suitable than nails. Usually there is no need for glue as well as nails, but a waterproof glue will strengthen the joint. Anything less than a fully-waterproof glue would have only a brief life.

For most assemblies, you can use common or box nails. If you want the heads to be less prominent or wish to punch them below the surface, casing or flooring nails with small heads are available. Their grip is not as strong so you need more of them. For increased hold in the lower piece of wood, ring-shank or barbed-ring nails are available. Another increased-hold nail has a twisted shank. For corrugated roofing, nails with special heads and other roofing nails made with large heads (FIG. 2-2) are available.

Driving a nail has a splitting action on the wood fibers. In most positions, this splitting is not enough to matter, but near an end or when using very large nails, it is advisable to drill before driving. You should drill deep enough to clear, or almost clear, the top piece, and you need to drill an undersized hole in the lower piece (FIG. 2-3A). An increased grip comes from driving nails at alternate slight angles (FIG. 2-3B) to give a dovetail effect. This technique of driving nails with increased grip should be used only when you are certain the joint will never have to be pulled apart. Levering the joint open then might break out fibers or split the wood.

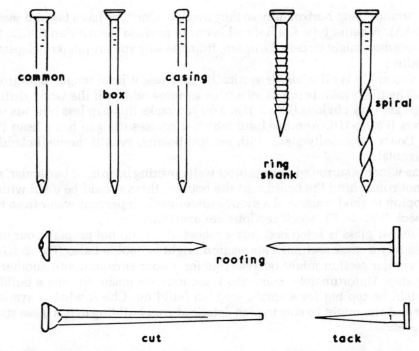

FIG. 2-2. *You use a variety of nails in constructing small wooden buildings.*

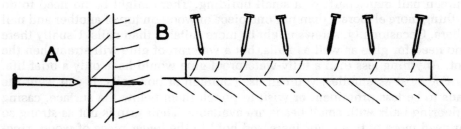

FIG. 2-3. *Drill for nails to avoid splitting (A). Dovetail nailing (B) is stronger than straight nailing.*

Nails commonly are made of steel and they probably will be your choice for general construction of a building, but steel will rust. It might not rust enough to matter, but nails in very wet conditions can rust away completely. Rust also might come through the paint and leave brown spots or streaks. You can purchase steel nails that are protected by *galvanizing* (coating with zinc) or other corrosion-resistant metals. Stainless-steel nails are expensive. Aluminum nails will not corrode enough to matter. If you are nailing through metal, it is better to use nails of the same metal, to prevent electrolytic action between different metals, causing corrosion. This corrosion might happen with aluminum guttering.

SCREWS AND BOLTS

Common, flat-head, wooden screws are the alternatives to nails in some circumstances. Other types might be more appropriate to hinges or other metal attachments. In some buildings, there is a need for very large screws. You then need a lag or coach screw, which has a head to suit a wrench (FIG. 2-4A). Drill for it and start it with a blow from a hammer.

Screws are made in at least as many different metals as nails and you can buy them with various protective coatings. Galvanizing tends to be rough, which might be an advantage because of its increased grip, if you are dealing with large sections of rough wood, such as posts.

Bolts are obtainable in many forms and sizes. In general, if you ask for a bolt, it is threaded to take a nut only part of its length. If you want it threaded almost to the head, you must ask for a screw. General-purpose-machine bolts have square or hexagonal heads and nuts (FIG. 2-4B). Stove bolts are long and thin bolts with screwdriver heads (FIG. 2-4C).

The bolts most useful in small buildings are coach or carriage bolts (FIG. 2-4D). Under the shallow-domed head is a square neck which pulls into the wood and prevents the bolt from turning as you tighten the nut. It also retains the bolt in the wood, although you might knock it out. This characteristic is valuable

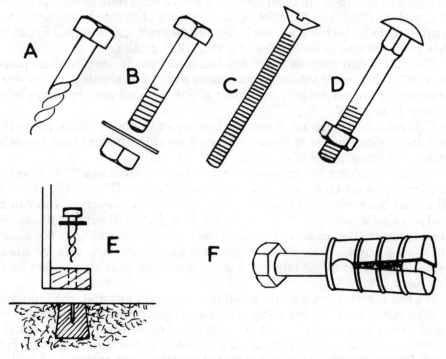

FIG. 2-4. You need large screws and bolts for joining parts of buildings and anchoring them to their foundations.

if you want to disassemble occasionally. Always use a large washer under nuts on softwood.

For attaching wood to masonry, you have a choice of method. This choice is particularly true on the foundations. You can set special bolts in the concrete as you lay it (FIG. 1-4C). An alternative is an ordinary bolt with a large washer under its head, to grip the concrete. The other way is to drill downwards, using a fastener that will grip the concrete. With the first method, you have to carefully locate the wood over bolt ends already there. In the second method, you can drill downwards through the wood and exact positioning is not so important.

Sometimes wooden blocks are let into the concrete (FIG. 2-4E), but they should be rot-resistant wood or you might find that the attachment to the foundation is negligible if the wood is weakened by rot.

Methods of attaching downward into concrete involve drilling fairly large holes, for which you will need special equipment. Several types of anchor are available. They have a part which goes into the hole, and it is then expanded by driving a bolt or screw into it. A typical anchor has a body in two parts. When you drive a bolt in, you force the parts outward (FIG. 2-4F).

JOINTS

You might be able to simply nail some framework parts from outside at a corner or T junction (FIG. 2-5A, B). If the opposite side is obstructed, you might have to nail diagonally inside (FIG. 2-5C). To ensure exact location, you could nail guide pieces at one or both sides of a joint (FIG. 2-5D).

Depending only on nails is not very reliable for positioning. Shallow rabbets (FIG. 2-5E) will prevent sideways movement without weakening the cut wood. These rabbets are particularly important at window and door openings where accuracy is essential.

Corners could have open mortise-and-tenon joints or bridle joints (FIG. 2-5F). In window frames or similar places, it would be better to use haunched mortise-and-tenon joints (FIG. 2-5G).

You can reinforce any corners with metal or wooden gussets. Also you can nail pieces of sheet metal at one or both sides of a joint (FIG. 2-5H). The metal will not be thick enough to matter under any outside covering. You can use wood in a similar way (FIG. 2-5J) on the inside if there will not be a lining. You can use very similar gussets in parts of the roof. If the building is big enough to have roof trusses, their joints might have gussets on each side. At the top of a gable, a gusset might join the framing parts and provide a socket for the ridge piece (FIG. 2-5K).

If a building is clad with plywood or other sheet material, the framework might be stiffer than any cut joints. Once you have covered a framework, you should have a very strong assembly. With board covering there is more risk of distortion and stronger frame joints are advisable. Further stiffening might come from a lining, but if it is only hardboard or other thin material, it will not contribute much strength.

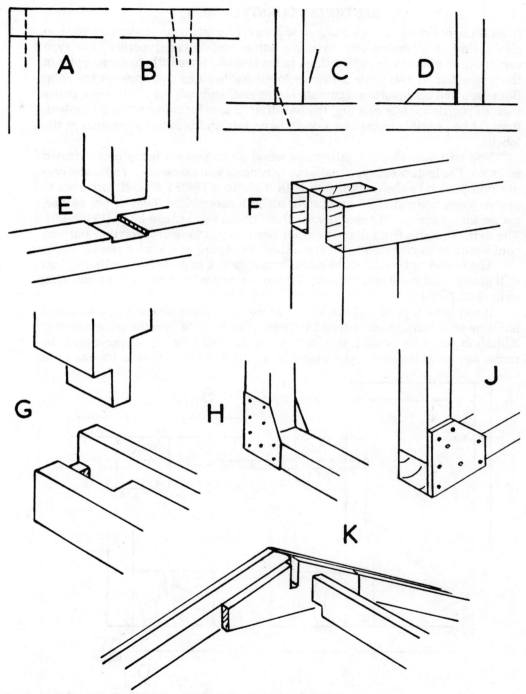

FIG. 2-5. Nail joints in framing or hold them with cut joints.

SECTIONAL CONSTRUCTION

Even if the building you are making will never be taken apart and moved, you might find it advantageous to make some prefabricated parts. This type construction allows you to take them to the site and have little to do except join them together. If you have the space to assemble sides and ends on the shop floor, you can do squaring, accurate laying out, and cutting joints more easily than in position while erecting the building. If size for transport is a problem, it might be possible to make a side in two parts and bolt them together at the jobsite.

You will have to use a little more wood when you are using prefabricated sections. For instance, if you make everything as you assemble on site, one corner post will take the covering in both directions (FIG. 2-6A). If you wish to prefabricate, there must be an upright on both assemblies. They might nail together for a permanent assembly, or there could be carriage bolts (FIG. 2-6B). The extra strength from double corner posts might be worth having, but you could reduce sections and still have them as strong as a single post.

The covering boards could merely overlap at a corner (FIG. 2-6C), but you will protect end grain and improve the appearance when you nail in a covering strip (FIG. 2-6D).

If you have framed and boarded the roof, you might attach it in a sectional building with carriage bolts, if you arrange strips to come over the gable framing. Although you might drill holes for bolts in the roof framing, or position bolts under the outer boarding, you might leave drilling the gables or trusses until

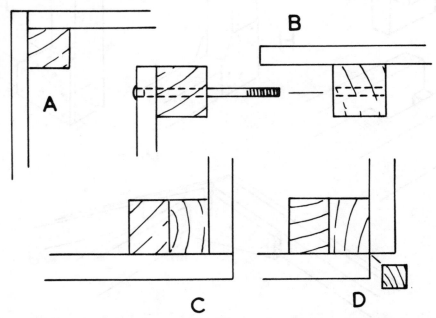

FIG. 2-6. For the simplest corners, nail them on one post. Others might bolt through two posts, with the skin overlapping or with a filler in the corner.

the first assembly, to allow for slight variations, particularly with a ridged roof. Subsequent disassembly and reerection will be helped if you mark all meeting parts.

When prefabricating, it is easy to accurately square door and window openings and make the fitting parts, but you might have to plane for accurate fitting after the building is in position. Even if you intend a door to reach the floor, it is advisable to make the side or end which will contain it with the bottom strip going across the opening, to keep the assembly free from twist. Cut through it after you attach the wood on each side to the floor or foundation.

3

SIMPLE

STORAGE UNITS

SEVERAL SMALL BUILDINGS, OR STRUCTURES SOMETIMES ALMOST TOO SMALL TO BE called buildings, might be useful on your property. A common need is for storage of garden tools, possibly some distance from your garage or main storage place, if your garden is extensive. Other items might not need much space, such as loose boating equipment near a dock or barbecue items that you prefer to keep apart from other things.

Construction can be similar to any other small building, although some parts might be smaller and the work simpler. If your experience of constructing wooden buildings is slight or you are a beginning woodworker, one of these smaller assemblies will make a good introduction. It is unlikely you will go wrong, but if you do, you will not waste much material.

A very compact assembly is attractive, but make sure it will hold all you expect it to, and probably a little more. If you are preparing for garden hand tools, you do not want to have to force in something with a long handle diagonally, when you can plan for a few more inches to allow it to go in straight and cause less of an obstruction to other tools. Gather the equipment you plan to store and check to make sure you are allowing adequate internal measurements.

UPRIGHT GARDEN TOOL LOCKER

You can make a storage unit quite simply with framed plywood. The locker shown in FIG. 3-1 is made mostly from ½-inch-exterior plywood on 1½-inch-square strips. It could fit against a fence or stand as an independent unit. As shown, it is assumed that it has a dirt floor, but a plywood bottom could rest on the lower framing. The door lifts out, which allows the maximum access to the inside, but you could hinge it at one side.

Check the sizes of your equipment, but the locker shown in FIG. 3-2 should hold most gardening hand tools and lighter power tools.

Make the pair of sides first (FIG. 3-3A). Cut the plywood to shape and nail it to the framing strips. There is no need to cut joints between the framing parts.

FIG. 3-1. *This tool locker stores most tools upright.*

If you wish, allow extensions at the bottom (FIG. 3-3B) to go into the ground. The back (FIG. 3-3C) goes over the sides and has stiffening pieces at top and bottom. Plane the top-ply edge and its strip to match the shape of the roof.

At the front strips, go across above and below the door (FIG. 3-3D,E). They could be solid wood or plywood. Inside each put stiffening strips.

Check squareness in all directions, then add the roof. If the locker will come against a wall, stop the roof level with the back, but otherwise you can overlap it a small amount. Allow a good overlap on the other edges and around all corners.

Make the door to overlap the sides and fit easily between the front strips (FIG. 3-3F). Put a stiffening strip inside the top and one at the bottom to hook over the front piece (FIG. 3-3G). These strips should fit easily between the locker sides so you can put the door in or take it out without trouble.

You could drill two finger holes at the top of the door or cut a wooden block as a handle. Make two strip-wood turnbuttons to hold the door closed. Alternatively, fit a hasp and staple for a padlock or add an ordinary door lock.

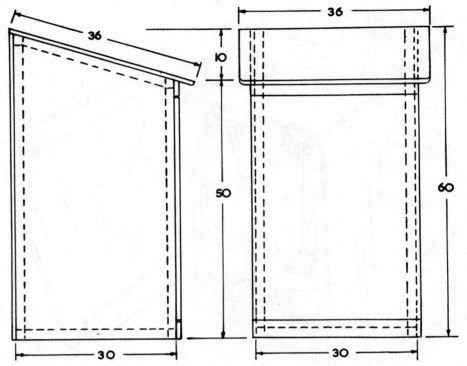

FIG. 3-2. Sizes of the upright garden-tool locker.

Materials List for Upright Garden Tool Locker

2 sides	½- × -30- × -60 plywood
1 back	½- × -31- × -60 plywood
1 top	½- × -36- × -36 plywood
1 door	½- × -31- × -44 plywood
2 frames	1½ × 1½ × 60
2 frames	1½ × 1½ × 52
8 frames	1½ × 1½ × 30
2 front strips	½ × 3 × 32
2 door strips	½ × 3 × 29

HORIZONTAL GARDEN TOOL LOCKER

If there is no convenient wall to put an upright locker against or you want something less prominent to store tools, a horizontal locker might be preferable. If you paint it green or allow it to weather to a natural color, it will be inconspicuous. The locker shown in FIG. 3-4 has a sloping lid to shed water and part of the front opens for easy access to the tools inside. The bottom is raised clear of the ground.

You can use solid wood throughout, but these instructions assume the skin is made of ¾-inch-exterior plywood, with 1½- or 2-inch-square framing. As with

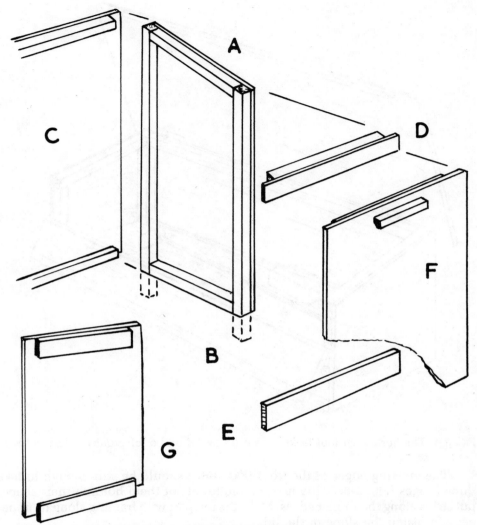

FIG. 3-3. Constructional details of the upright garden-tool locker.

the first locker, check the space you need for your tools. The sizes suggested should suit most needs (FIG. 3-5).

Make the pair of ends (FIG. 3-5A and 3-6A). Nail the plywood to the framing, preferably with the crosswise pieces overlapping the uprights. The back (FIG. 3-6B) overlaps the ends and you should stiffen it at the top and the bottom with pieces which fit closely between the ends.

Form the front by hinging two pieces (FIG. 3-5B) together. Make the lower part with a stiffened bottom edge, in the same way as the back. Before nailing the lower part on, make the bottom (FIG. 3-6C) to rest on the four bottom framing pieces, with notches around the uprights. Nail in the bottom and the front.

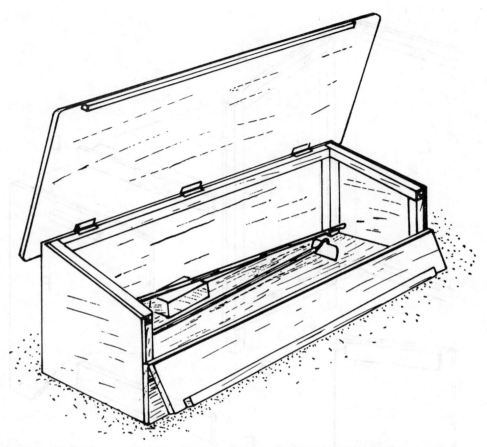

FIG. 3-4. *The horizontal tool locker has a lifting lid and a fall front for easy access.*

The meeting edges of the front (FIG. 3-6D) should be stiff enough to take three hinges (FIG. 3-5C). If you are doubtful about them holding their shape, put strips along their inner edges. Make the top part with framing along its upper edge to match the slope of the lid.

Make the lid with about 1½ inch overlap all around it. Round its corners and edges. If it is necessary to stiffen the lid, put pieces across inside so they clear the framing when you close it. The front strip (FIG. 3-5D) should hold that edge in shape. Hinge the lid to the back. Be sure it rests on the ends and the flap at the front. If necessary, plane these parts for a reasonably close fit, then there should be no need for fasteners or handles. The weight of the lid will keep the flap closed, but a slight lift at the front allows you to pull open both hinged parts.

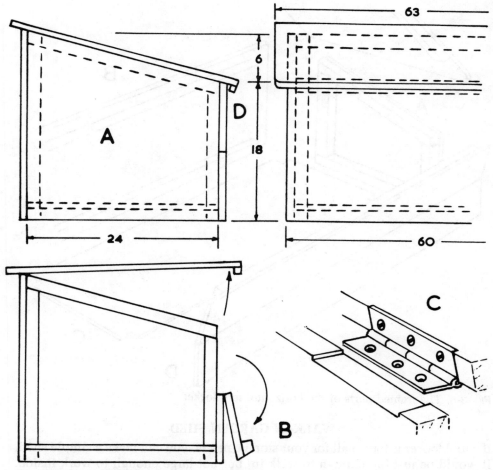

FIG. 3-5. Sizes of the horizontal tool locker.

Materials List for Horizontal Garden Tool Locker

2 ends	¾- × -24-	× -24 plywood
1 back	¾- × -24-	× -60 plywood
2 fronts	¾- × - 9-	× -60 plywood
1 lid	¾- × -30-	× -65 plywood
1 bottom	¾- × -24-	× -60 plywood
6 frames	1½ × 1½ × 25	
2 frames	1½ × 1½ × 18	
4 frames	1½ × 1½ × 60	
1 lid edge	¾ × ¾ × 60	

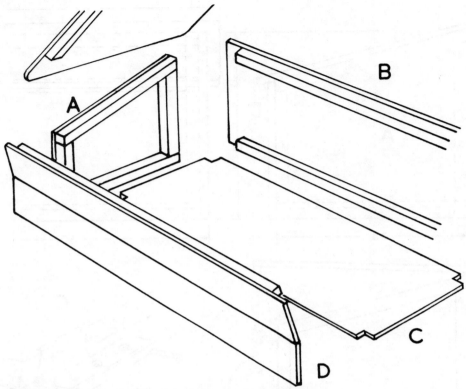

FIG. 3-6. The framed parts of the horizontal tool locker.

WALK-IN GARDEN SHED

If a tool locker is too small for your storage needs, you can make a simple shed. It would be just big enough to walk in, but not large enough to work inside. It need not have windows and the headroom might be minimal. If these are your needs, the shed in FIG. 3-7 is easy to make and the materials cost much less than for a larger building.

The suggested sizes shown in FIG. 3-8A are for a sectional building that you can make elsewhere and assemble in position, or you can disassemble it if you wish to move it. You can use plywood completely for the cladding, or you can board the walls and the roof or make the roof of plywood. In either case, cover the roof with roofing felt or other waterproof material. Framing might be nearly all 2-inch-square sections. Board the door or make it of plywood.

The pair of ends control other sizes. Allow for the door in one end (FIG. 3-8B and 3-9A). At the other end, the upright might be central. If the covering is plywood, it might provide sufficient strength at the corners for the framing pieces to abut against each other, but otherwise, you should join them in one of the ways described in Chapter 2. Locate all joints away from the corners with notches (FIG. 3-8C). Make sure the doorway will finish square. Horizontal boarding is shown in FIG. 3-8, but you could nail on vertical boards. At the high

FIG. 3-7. *The walk-in garden shed allows you to get into the storage unit.*

side of the door panel put a vertical board (FIG. 3-8D). Finish the boards or plywood covering level with the framing all around.

Frame the back and front in the same way, with corner and intermediate joints similar to those of the ends. Use the ends as a guide to heights, and bevel the top members of both panels to match the slope of the ends. Two intermediate uprights (FIG. 3-9B) should provide sufficient stiffness.

Allow for the boards on the back and the front to overlap the ends, so the end uprights are set back sufficiently (FIG. 3-8E). Drill for ⅜-inch bolts. A bolt near the top and the bottom and two spaced between the top and the bottom should be enough on each corner.

For a boarded door (FIG. 3-10A), three ledges should be level at the hinge side, but set back a little at the other side. One brace should be enough to prevent sagging. Position the top and bottom ledges so you can screw into them from the hinges, which might be T hinges on the surface, or you can let in ordinary hinges between the door and its post. A metal loop handle would be suitable

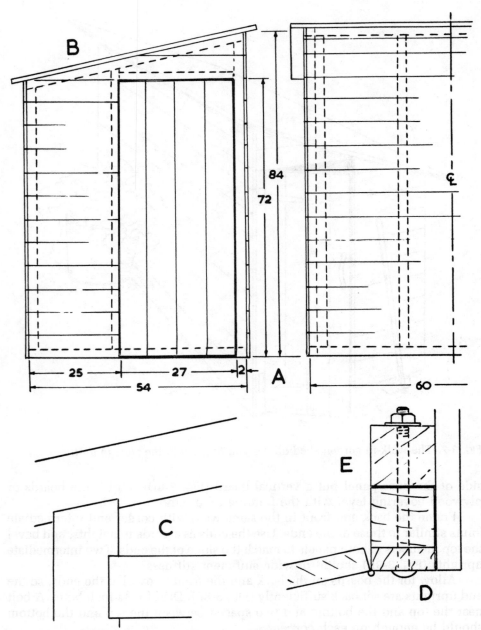

FIG. 3-8. Sizes and structural details of the walk-in garden shed.

or you could make one out of wood or turn a knob on a lathe. The simplest fastener is a strip-wood turnbutton, but you could fit a lock on the door. The door will close against the bottom framing strip. That strip might be sufficient, but you could put a short stop piece near the top as well.

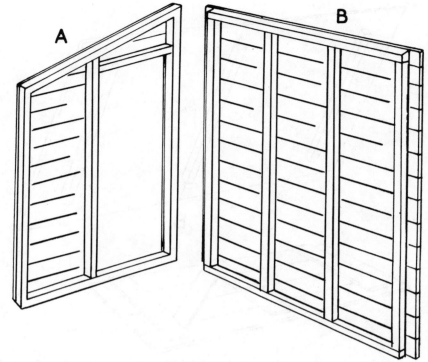

FIG. 3-9. Two walls of the walk-in garden shed.

Materials List for Walk-in Garden Shed

Ends

2 uprights	2 ×	2 × 86
4 uprights	2 ×	2 × 75
2 bottoms	2 ×	2 × 56
2 tops	2 ×	2 × 59
1 door rail	2 ×	2 × 30

Back

4 uprights	2 ×	2 × 75
2 rails	2 ×	2 × 60

Front

4 uprights	2 ×	2 × 86
2 rails	2 ×	2 × 60

Roof

3 strips	2 ×	2 × 54
2 panels	¾- × -35- × -66 plywood	

Wall covering could be ½-inch or ¾-inch plywood or ¾-inch boards, with the lengths needed depending on widths available.

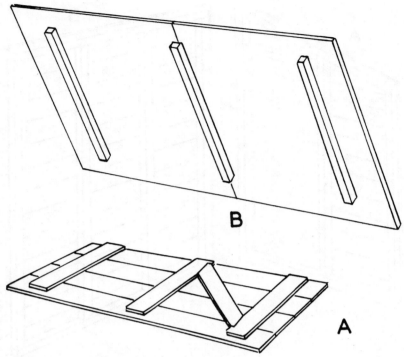

FIG. 3-10. Door and roof for the walk-in garden shed.

The roof is larger than you can make from a normal-size, single sheet of plywood. You might have a joint at the center (FIG. 3-10B). Make the size sufficient to give an overhang of about 3 inches all around. Three strips across should fit easily between the shed back and front. Those at the ends come inside the shed ends. They will hold the building square and you should bolt them through in a similar way to the corners—three bolts on each end should be sufficient.

Although you might let the plywood roof with just a paint finish, it will be better if you cover it with a waterproof material. This covering should be the last job after you erect the building, bolt it down to its foundation, paint all the woodwork or treat it with preservative. For the best protection, paint inside the corners before bolting them together.

LEAN-TO SHED

If there is an existing high wall, fence, or side of a house, you might wish to make a small building as a lean-to attached to it. This construction means you already have a rigid support that will hold the building in shape and protect it from the effect of high winds. If it is upright and straight, it gives you a datum for squaring the building you are adding. If you check and find it is not true, do not make the new building to conform to its errors. The added shed should be upright and square or it will look wrong. Where the new shed meets the untrue wall, you should fair it in. You might be able to cut covering boards to conform or you might have to attach shaped uprights to the wall, with their outer edges vertical. Treat the roof joint similarly. Obviously, the roof joint must be weathertight.

Consider the effect of the addition on its surroundings. Will you have to divert a path around it? Will it obstruct light or view through an existing window? When you put down foundations, of whatever type, make sure they cannot cause rainwater to run back into the house foundation. If you are joining onto a fence, is it in suitable condition for your purpose? You might be giving yourself more work in the future if you will have to replace the fence because of rot or for other reasons.

If you want to walk into it, a lean-to building must have enough height for a door. Slope the roof away from the existing wall. You can have a flat roof, but you should give it a slight slope so it sheds water. A steeper slope is usually better, particularly if you get heavy snowfalls. For a gardener's shed, the height at the outer wall might be less than 6 feet, if there is a worktop inside, so you do not walk right up to the wall. You might modify the sizes suggested to suit your situation.

The building shown in FIG. 3-11 is covered with vertical, matched boarding. For horizontal boarding, arrange the framing upright. For a plywood skin, the framing is satisfactory as shown. All of the parts are prefabricated—two ends and a front, with a roof that you can make in position.

The ends settle the sizes of other parts. Make the door end first (FIG. 3-12). It is shown 72 inches wide with a height sloping from 90 inches to 72 inches, but this stage is where you make your modifications, if required. Assemble the frame with 2-inch-square strips, notching them into each other, and jointing the corners or using gussets (FIG. 2-5). The strip at the outer corner (FIG. 3-12A) need only be 1 inch thick to fit inside the front piece.

At top and bottom of the high edge, notch the framing to take boards which you will use to screw the assembly to the supporting wall (FIG. 3-12B,C). Cover the framing with tongue-and-groove boards which have about a ¾-inch-finished thickness and which are 6 inches wide. Allow for the overlap on the front (FIG. 3-12E).

Fit strips around the door sides and top to cover the boards and framing (FIG. 3-12D). Let these project forward with rounded edges. Leave the frame strip across the bottom of the door opening, either permanently or until after you have fastened down the walls.

Use this first end as a pattern for making the opposite end identical. Carry the rails right across. The boarding probably will provide adequate vertical stiffness, but if necessary, include some uprights between the rails. Cover that end with tongue-and-groove boards. If you wish, you can add another door or a window in the end, arrange it as described for the front.

Make the door with upright boards similar to the covering. Put three ledges across and one or two diagonals. Put a strip as a stop on the side of the opening

FIG. 3-11. *This lean-to shed fits against a wall and has windows in its lower side.*

the door will close against. Arrange three hinges, a handle, and a lock or other fastener.

Make the front (FIG. 3-13) to match the height of the ends. You could bevel the top to match the slope of the roof, although you should get a close enough fit for most purposes by letting the roof rest on the front edge. Make all the framework with 2-inch-square strips, with joints similar to those in the ends. Corners fit inside the shed ends (FIG. 3-12F). Cover the framing with upright boards in the same way as the ends.

For the window openings, make a sill the full length, notch it around the intermediate uprights (FIG. 3-13A). Put pieces of covering board over the two uprights and frame around the window openings in a similar way to the doorway (FIG. 3-13B). Put strips around to make recesses for the glass (FIG. 13-3C).

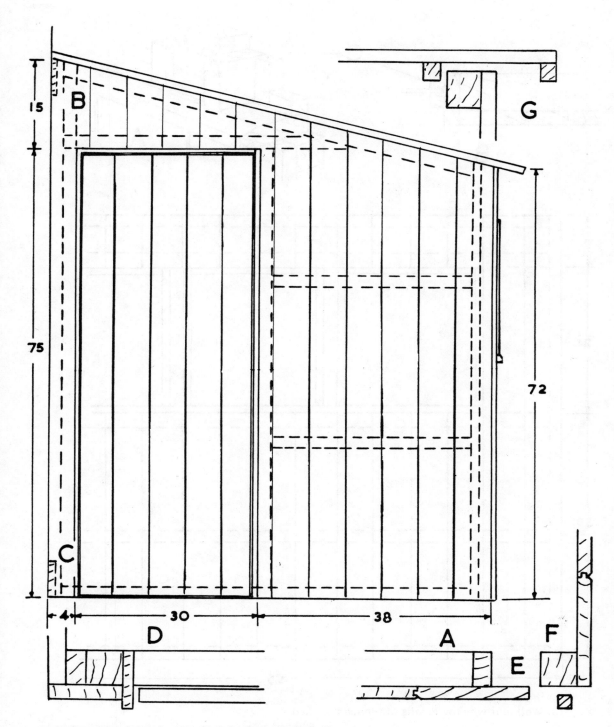

FIG. 3-12. Details of the door end of the lean-to shed.

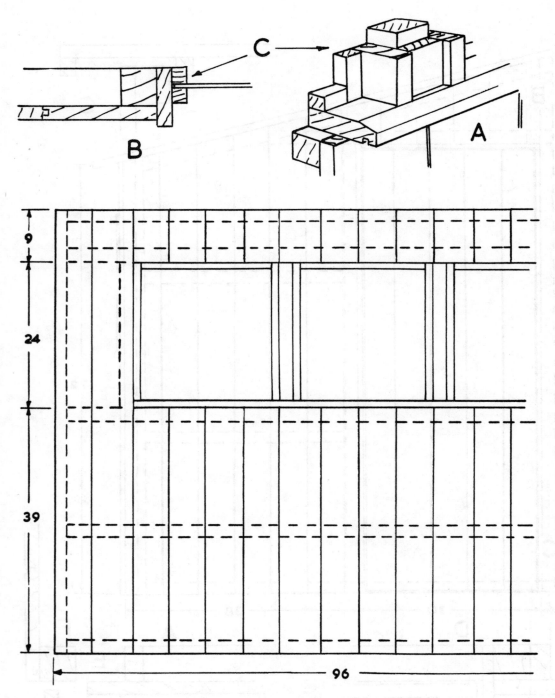

FIG. 3-13. Wall and window details of the lean-to shed.

Postpone fitting the glass until you have erected the shed. When you do fit it, putty the glass into the recesses or use more wooden strips.

As you erect the ends and front to the wall, screw or bolt the horizontal strips into their recesses at the back and fasten through them into the wall, taking care that their lengths match the front, or you will be unable to square the assembly. If you do not expect to move the building, you can nail or screw the front corners. Otherwise, use bolts. Fit square strips to cover the board edges.

The roof could be ¾-inch plywood or tongue-and-groove boards laid with their smooth sides upwards. In both cases, use framing strips at about 18 inch intervals to fit inside the walls. Allow for an overlap of about 6 inches on the shed walls. At the supporting wall, bevel to make a close fit and fair the edges if necessary. Edge the outer limits of the roof (FIG. 3-12G). More strips inside will help when locating places where other roof framing parts do not come.

Nail or screw on the roof. Cover it with tarred felt or other material. At the edges, wrap around the border pieces. At the top edge, the joint will depend on the material of the fence or wall. If possible, carry the covering material up so existing clapboards or other projections will shed water over the cover rather than under it. Screw strips through the covering to secure it. It might be necessary to use a jointing compound or mastic to ensure complete waterproofing.

For storing and preparing garden plants and equipment, it might be sufficient to paint the building and leave it as it is. For a neater inside and to provide some insulation, line the walls with hardboard or plywood. A broad shelf the full length under the windows will serve as a bench and also will brace the walls. Other shelves and storage arrangements attached to two or three walls also will provide stiffness.

Materials List for Lean-to Shed

4 uprights	2	×	2	×	92
6 uprights	2	×	2	×	74
2 uprights	1	×	2	×	74
5 rails	2	×	2	×	96
2 rails	2	×	2	×	76
2 door rails	2	×	2	×	50
4 end rails	2	×	2	×	36
2 end rails	2	×	2	×	56
4 window posts	2	×	2	×	26
1 window sill	1	×	4	×	74
18 window frames	¾	×	¾	×	24
9 window frames	¾	×	3½	×	24
2 door frames	¾	×	3½	×	76
1 door frame	¾	×	3½	×	32
2 backs	1	×	6	×	100
4 roof frames	2	×	2	×	96
3 roof edges	1	×	1	×	108

Covering—tongue-and-groove ¾ × 6

STRESSED-SKIN WORKSHOP

Plywood has considerable strength. Obviously it is stable compared with a covering of many independent boards. It holds framing and other parts in shape and is strong in itself. Much more strength comes from the skin than does an assembly with other covering. This fact applies if the plywood is on one side only, but if it is on both sides, a very rigid structure results. Even if there is only air between the skins, that provides some insulation. If you fill spaces with insulating batting or other materials, a good, all-weather building results.

The building shown in FIG. 3-14 is intended to have plywood inside as well as outside. Its sizes are arranged to cut standard sheets of 4-×-8-feet plywood economically. Two windows are suggested, but you can arrange others to suit your needs. The end door is far enough from one side to allow wide shelves there or a bench for a gardener's use. A bench for a hobby could go under the end window. If this is to be a year-round workshop a wooden floor is advisable.

Overall sizes are shown in FIG. 3-15A. For sides and ends, use the plywood sheets vertically. Arrange roof sheets similarly, with joints halfway along the roof. All of the framing is 2-inch-square wood.

FIG. 3-14. This small workshop has a skin made of plywood.

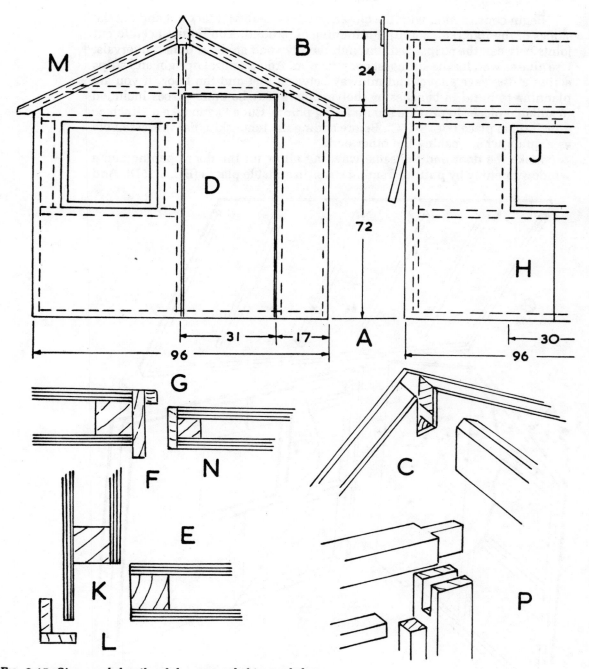

FIG. 3-15. Sizes and details of the stressed-skin workshop.

Begin construction with the closed end (FIG. 3-16A). Mark out and cut the slopes of the roof (FIG. 3-15B). Fit framing all around. You do not need to cut joints between the strips. Nail through the plywood at about 6 inch intervals. Use glue as well for the strongest construction. Add a central upright and strips across at the eave's level and midway between that and the floor. If you are planning to build in benches or shelves, there may be other pieces included in suitable places to provide secure fixing points. Cut away to take a 2-inch × 3-inch ridge piece (FIG. 3-15C). Before adding the inner skin, use this assembly as a guide when making the other end.

Make the door end the same way, but allow for the door opening and a window opening by putting framing strips in suitable places (FIG. 3-15D). Add

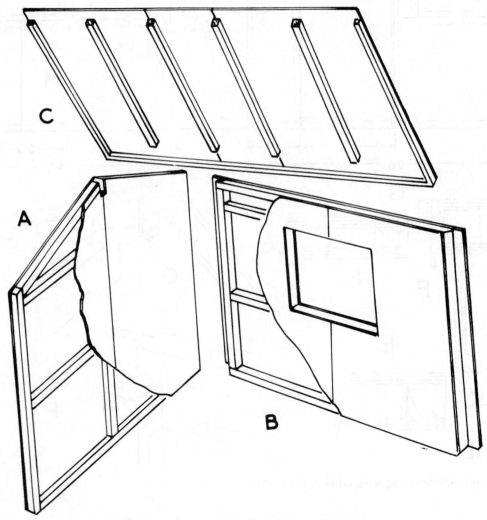

FIG. 3-16. Wall and roof details of the stressed-skin workshop.

additional framing where needed for shelves or benches. Arrange the bottom of the window to come a few inches above your bench and allow headroom in the doorway by keeping the cross member above eaves level.

When you are satisfied with the framing of both ends and the inner plywood, level the edges (FIG. 3-15E). At window and door openings, use strips to cover the plywood and extend them forward (FIG. 3-15F). Include strips at the inner edges to act as stops and provide draughtproofing (FIG. 3-15G).

The sides (FIG. 3-15H and 3-16B) are plain, rectangular assemblies on two pairs of upright sheets of plywood. Mark the height from the ends and bevel the tops to match the roof slope.

For a plain side, framing the edges should provide sufficient strength. Use a strip as a cover piece centrally and put a central, horizontal strip the other way. You might need other strips to provide places for benches or shelves. It is a good idea to put a window, the same size and height as the one in the end, in one side (FIG. 3-15J). Arrange horizontal framing to suit, with suitable

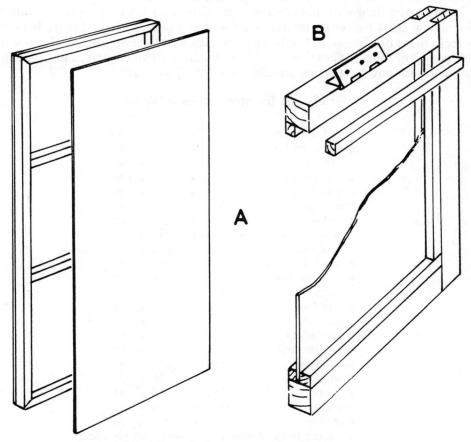

FIG. 3-17. Door and window construction for the stressed-skin workshop.

uprights. Cover the sides inside and out in the same way as the ends, except at the corners allow the outer plywood to extend over the end assemblies (FIG. 3-15K). When you put the building together, the corners will look neat and you will protect the edge-grain of the plywood if you nail on thin, wooden strips (FIG. 3-15L).

If you make the roof to finish flush with the ends, you will only need central joints. It will look better if it extends about 6 inches at each end and if you use sheets 48 inches wide; there will have to be pieces 12 inches wide as well. Allow a 6 inch overhang at the eaves.

Erect the building without its roof, and cut the top plywood to suit. There should be sufficient stiffness from the edge supports, plus intermediate strips (FIG. 3-16C). Put strips around the projecting edges. Cover with tarred felt or other material, wrapping it over the edges and nailing underneath. You can improve appearance by nailing on bargeboards. Arrange them to stand slightly above the roof surface and cut off the lower corners (FIG. 3-15M). Upright, pointed pieces at the center will finish the decoration.

Make the door with plywood on both sides and 1¼-inch-square framing. Include at least two cross members for stiffness (FIG. 3-17A). If there is to be a lock, put a block inside. Do the same to take screws from hinges, if you will be using T or other hinges that extend. Make the door an easy fit in its opening, allowing for covering strips around the edges (FIG. 3-15N).

Materials List for Stressed-Skin Workshop

2 end uprights	2	×	2	×	96
5 end uprights	2	×	2	×	74
5 end rails	2	×	2	×	96
5 side uprights	2	×	2	×	74
7 side rails	2	×	2	×	96
7 short rails	2	×	2	×	48
4 end slopes	2	×	2	×	60
2 door frames	¾	×	3½	×	75
1 door frame	¾	×	3½	×	32
8 window frames	¾	×	3½	×	30
8 window frames	⅝	×	⅝	×	30
1 ridge	2	×	3	×	96
8 roof frames	2	×	2	×	52
2 roof edges	1	×	1	×	96
4 roof edges	1	×	1	×	60
4 bargeboards	1	×	5	×	60
8 windows	1¼	×	1¼	×	30
8 windows	⅝	×	⅝	×	30
2 doors	1¼	×	1¼	×	75
4 doors	1¼	×	1¼	×	30
8 corners	½	×	2	×	72

Covering from 22 sheets ½- × -48- × -96 plywood

For fixed windows, you can put glass against the inner strips and either putty to them or add other strips, as described for the last project. Alternatively, you can make one or both windows to open. In any case, it is better to have a wider, sloping sill at the bottom, to shed water (FIG. 3-13A).

You can make a window which opens with molding and mortise-and-tenon joints, in the same way as for a house. You can simplify the construction for this shed, however, by using bridle joints and separate strips to make rabbets for the glass (FIG. 3-15P and 3-17B). Use putty or second strips to hold the glass. Hinge each window at the top and have a locking strut at the bottom.

Exterior-grade plywood should have quite a long life if left untreated, but cover the edges to prevent the entry of water, preferably by bedding the covering strips in waterproof glue. You will improve appearance and increase durability if you thoroughly paint the building. A green finish under a black roof and bargeboards should look smart and blend in with most surroundings.

CURVED-ROOF PLYWOOD UNIT

Plywood has considerable strength in itself, so a building made from it needs less framing as the panels contribute plenty of stiffness. The standard 4- × -8-feet-plywood sheets are large enough for you to complete your project with few joints. Of course, any plywood which will be exposed to the weather should be exterior grade. The quality plywood you choose depends on its purpose.

You can make a covered unit with a few sheets. The simplest unit would be with a sloping roof, but you will obtain an increase in strength and rigidity by curving the plywood. You can make the assembly in FIG. 3-18 from four sheets of ½ inch, or thinner, plywood, with framing of 2-inch-square strips. Of course, you will not be able to walk into this building as the doorway is only 27 inches high. You could use it for sheltering tools, equipment, or for animals. If you want to treat it like a small tent, it would be possible to sleep inside!

The roof is a full sheet without cuts. The front and back are full width, but reduced in length. The ends are made from half sheets. You can complete construction quick if you need the unit in a hurry. Making this shelter is probably quicker than any other building in this book. If you are able to spend more time on it, you can give it a better finish.

Mark out the front (FIG. 3-19A). Cut the piece to length and mark a centerline on it. You can obtain the curve of the top by bending a batten around and penciling against it, but you will get a better curve, which is part of a circle, by using an improvised compass. Have a strip of wood just over ll feet long. At 11 feet from one end, push through an awl. Extend the centerline from the sheet and position the awl in the floor on this line so the end of the "compass" is on the edge of the center of the sheet. Pull this "compass" around to draw the curve with a pencil against the end. Cut the curve and use the front to mark a matching back.

When you have assembled the parts, remember to place a bevelled strip across the end to take the roof (FIG. 3-19B,C). Mark where this strip goes on the front, using the actual piece of wood as a guide to size. Put a strip of wood across

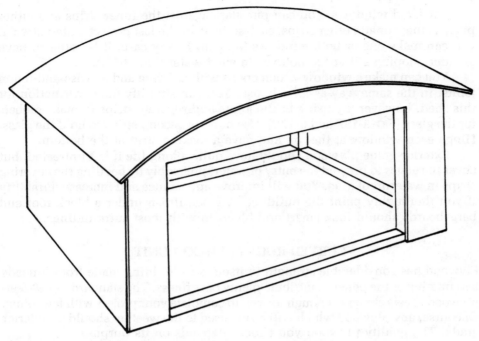

FIG. 3-18. A curved-roof unit made from sheets of plywood.

immediately below this beveled strip and mark the height of the doorway (FIG. 3-19D,E). Cut out the plywood to this height and 12 inches in from the ends (FIG. 3-19F).

Put a rail across the bottom of the opening and another across the top, with uprights at the ends and at each edge of the doorway. You can nail these parts through the plywood, but for a better construction, use waterproof glue as well.

At the back, put strips at the ends and across the bottom, leaving gaps at the top corners for the bevelled strips that you will put across there. Cut the end plywood pieces to match the heights of the back and front and 42 inches across (FIG. 3-19G). Nail the ends to the back and front (FIG. 3-19H). Put bevelled pieces across the top edges and square pieces across the bottom edges.

Stand the assembly on a level surface and check for squareness. Put three more pieces across to hold the roof in shape (FIG. 3-19J), nailing them through the back and front and leveling them so their upper surfaces are level with the curved edges.

See that the assembly is still square, then bend the roof sheet around, with help if necessary, to check the amount of overlap at its sides and each end. Mark on it where the walls come as a guide to nailing. Start at one end to nail the roof sheet down. Progress from there, nailing to each of the crosspieces in turn, until you finally nail to the bevelled piece at the other end. It is the nails at the ends which are important. If you have very stiff plywood, it might be advisable to alternate screws with the nails at the ends.

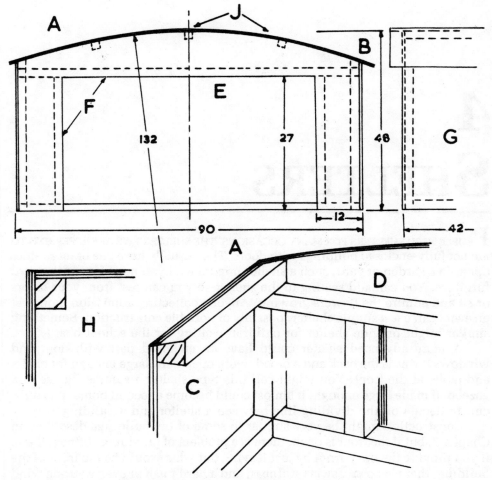

FIG. 3-19. *Size and construction of the curved-roof plywood unit.*

You should bevel or round the corners of the roof to prevent splintering. For a better shelter, you might take off sharp edges all around. As designed, the unit goes directly on the ground. You could nail a piece of plywood to the underside of the walls to make a floor, but plywood is a better floor placed on top of the bottom edge framing. The plywood will be easier to fit before you put the roof on.

Materials List for Curved-roof Plywood Unit

3 sheets plywood	48	×	96	×	½
3 rails	2	×	2	×	92
6 uprights	2	×	2	×	30
7 rails	2	×	2	×	44

4
SHELTERS

P EOPLE NEED SHELTERS FOR MANY OCCASIONS. THE SHELTERS WE ARE REFERRING TO are not fully-enclosed buildings with doors. They usually have one or more sides open. In a garden or yard, such a shelter might be all you need for storing yard furniture. You can add seating to the shelter so you can rest from your labors or sit and admire the flowers growing. Anyone collecting admission money at an event can use a simple shelter, possibly of portable construction. Something similar might provide shelter for children waiting for the school bus.

A more advanced shelter could have an enclosed part with door and windows towards the back and a broad, more open part, large enough for chairs and table, at the front. You might call this type shelter a summer house or a gazebo. If made large enough, it almost could become a second home. It is difficult to decide on the dividing line between a shelter and a building.

Construction might be very similar to some of the buildings described in Chapter 3, but if the front is open, there is a problem of providing stiffness there. If you provide the open front by just leaving out what would be the front of the building, there is no crosswise stiffness, and a hard push or even a strong wind on one side could distort or even collapse the shelter.

TAKE-DOWN SHELTER

The take-down shelter is a basic, open-fronted shelter, which you can make permanent, but which is suitable for disassembling into flat sections by removing a few bolts. It has a wooden floor, which keeps the assembly square. As shown in FIG. 4-1, this structure is intended to be a one-person shelter for a ticket seller or money taker at an event. It would also provide shelter from a storm or a storage place for a few garden tools. The sizes suggested in FIG. 4-2A are for a shelter of this type, but you can use the same method for a shelter of many different sizes.

You can board the walls on 2-inch-square strip frames. The roof and floor might be ¾-inch plywood. Assemble the building with ⅜-inch coach bolts, which might be left in the outer parts when you take down the shelter.

FIG. 4-1. You can take apart this open-fronted shelter.

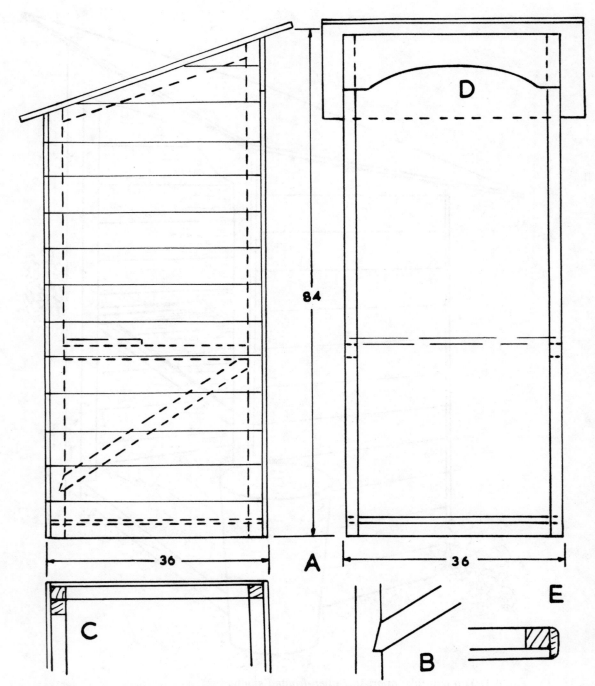

FIG. 4-2. Sizes and construction details of the take-down shelter.

Start with the pair of sides (FIG. 4-3A). The nailed boards might not be enough to keep the sides square, so include cross members and diagonals (FIG. 4-3B). If the cross pieces are about 30 inches from the ground, they can support a board across to act as a temporary high seat or somewhere to put boxes of tickets

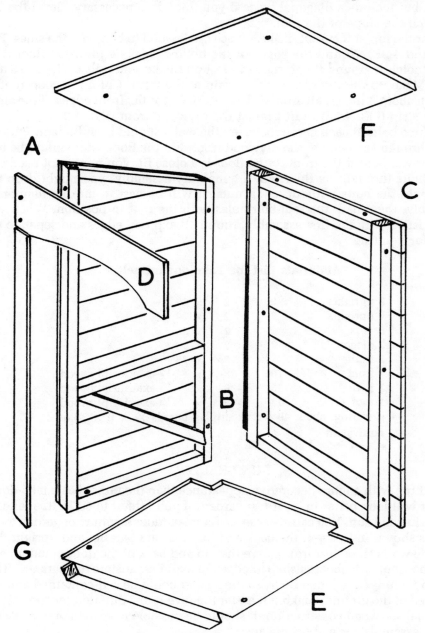

FIG. 4-3. Assembly of the parts of the take-down shelter.

or other items. So you do not interfere with fitting or removing the floor, notch the diagonals into the rear uprights a few inches up (FIG. 4-2B).

Make the back to fit between the sides (FIG. 4-2C), with the covering boards extending to overlap the side boards (FIG. 4-3C). This assembly should hold its shape, but include a diagonal brace if you think it is necessary. Bevel the top to match the slope of the roof.

For the top of the front, make a piece of plywood to bolt onto the sides (FIG. 4-2D and 4-3D). Round the edges of the hollow. Make a plywood floor (FIG. 4-3E); notch it around the uprights to rest on the bottom rails of the sides and back. Make up the thickness with a strip at the front. Cut a plywood roof to overlap about 3 inches all around (FIG. 4-3F). Cover the front edges of the sides with a strip (FIG. 4-2E) to fit against the plywood front (FIG. 4-3G).

Three bolts at each corner between the walls should be sufficient. Put two bolts through each end of the plywood front. For the floor, there might be two bolts upwards near the front. If the floor is a close fit, you might not need any other bolts through the floor. For a temporary assembly, you might drop the floor onto the bolts without using nuts. It will be best to make a temporary assembly, then mark the roof-bolt holes with the roof in position. When you are satisfied with the first assembly, round all exposed edges and separate the parts for painting.

Materials List for Take-down Shelter

2 uprights	2	×	2	×	86	
4 uprights	2	×	2	×	74	
8 rails	2	×	2	×	36	
2 diagonals	2	×	2	×	40	
2 fronts	¾	×	2¾	×	76	
1 front	10-		×-38-	×	-¾-exterior plywood	
1 floor	36-		×-36-	×	-¾-exterior plywood	
1 roof	42-		×-48-	×	-¾-exterior plywood	

Covering: about 40 pieces shiplap boards ¾ × 6 × 38 or equivalent

SUN SHELTER

Something with a more decorative appearance than the rather basic take-down shelter will look better in a yard or garden, if you intend to sit in it, sheltered from wind and sun. You can also use it to store outdoor furniture or garden tools.

As shown in FIG. 4-4, the back and the front are parallel and upright, but the side walls slope inwards. Curve the top and back of the door opening, and give the front bargeboards shaped edges to avoid an austere appearance. This design includes a wooden floor, so the shelter could be self-contained and not attached to the ground, making it easier to move it to a different location. With the usual softwood construction, it should be possible for two men to carry the whole assembly for a short distance.

FIG. 4-4. *The sun shelter has sloping sides and decorative bargeboards.*

The sizes suggested in FIG. 4-5A allow for occupying a ground area about 48 inches × 60 inches, but you can modify the sizes to suit your needs, providing you do not increase the size excessively. The skin suggested is shiplap boarding, but you can use plywood or other sheet material. You can use boards or plywood for the roof, then cover it with roofing felt or other material. Nail the bargeboards on after you have covered the roof. You can build in any seating or you can rely on separate chairs.

The key assembly is the back (FIG. 4-6A). Set this part of the building out symmetrically about a centerline. With the usual covering, one central upright should be all that you need to supplement the outside framing, which you can halve together (FIG. 4-6B). Cover the back with boarding, working from the bottom up.

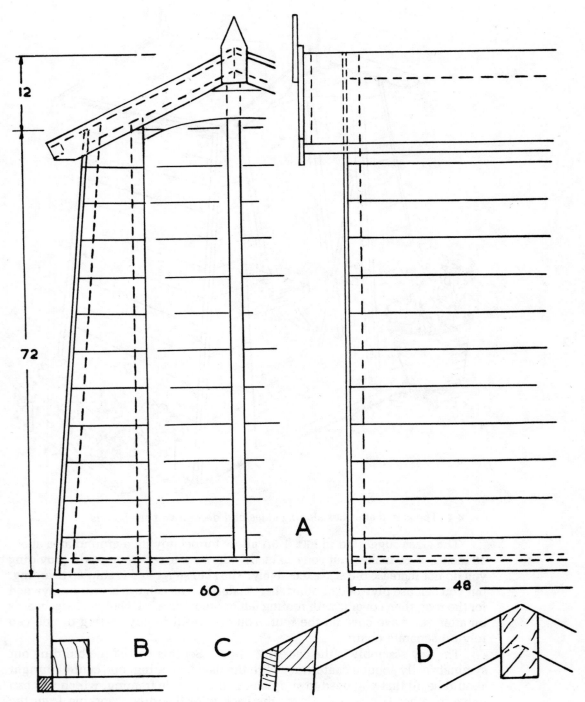

FIG. 4-5. Sizes and corner joints of the sun shelter.

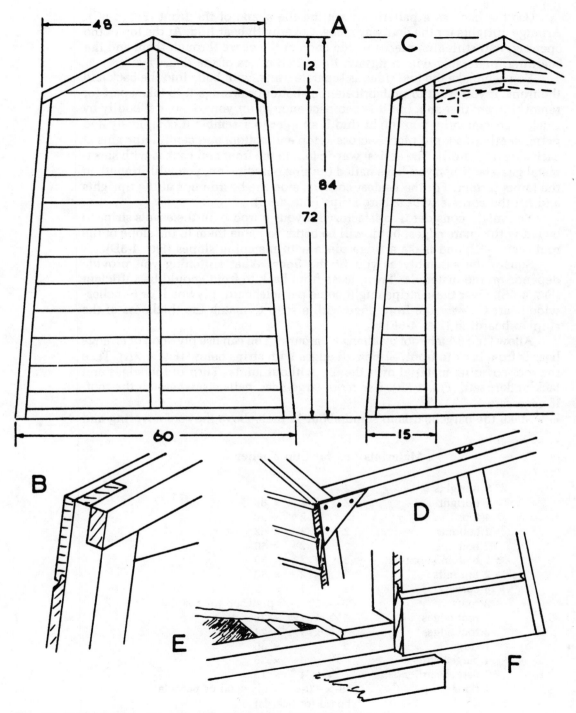

FIG. 4-6. Back, front, and constructional details of the sun shelter.

Use the back as a pattern for getting the shape of the front (FIG. 4-6C). Arrange uprights for the doorway sides. Cover with boarding. At the top of the opening, nail stiffening pieces inside and cut the curve through them and the boarding, preferably with a jigsaw. Round all edges of the doorway.

You can make the two sides as separate units, with bolts into the back and the front, if you want to prefabricate the shelter or arrange it to take apart for removal to another site, but it is compact enough for you to move it bodily by truck. Consequently, you might find it simpler to assemble it completely and permanently. If so, put pieces across at top and bottom and central uprights at each side (FIG. 4-6D). The easiest way to join to the front and back is with sheet-metal gussets. When you have nailed the side boarding on, you will strengthen the joints further. For the neatest corners, stop the board ends at the uprights and fill the corners with square strips (FIG. 4-5B).

You might consider it satisfactory for your purpose for the eave's strips to be left with square edges, but it will be better to plane them to the slope of the roof (FIG. 4-5C) and make a ridge piece with matching slopes (FIG. 4-5D).

Square the assembly as you fit the floor. What stiffening you provide depends on the materials. Two pieces from back to front should be sufficient (FIG. 4-6E). Over these strips might come particleboard, plywood, or 6-inches-wide boards. Cover the front edge with a strip to match the thickness of the shiplap boarding (FIG. 4-6F).

Allow for a 6-inch roof overhang all around. You can use plywood or arrange boards from back to front. Stiffen all edges with strips below (FIG. 4-7A). Take the roof-covering material over the top without joints. Turn in the edges and tack underneath (FIG. 4-7B). Use more large-head nails elsewhere on the roof, if necessary.

Make the bargeboards to stand about ½ inch above the roof covering and

Materials List for Sun Shelter

8 uprights	2	× 2	× 75
1 upright	2	× 2	× 86
4 tops	2	× 2	× 30
2 bottoms	2	× 2	× 62
2 bottoms	2	× 2	× 50
2 bottom supports	2	× 2	× 50
2 top rails	2	× 2	× 50
4 corners	1	× 1	× 75
2 roofs	36- × -72- × -¾ plywood (or boards)		
4 roof edges	1½ × 1½ × 36		
2 roof edges	1½ × 1½ × 74		
1 ridge	2	× 4	× 50
4 bargeboards	¾ × 5	× 36	
2 bargeboard ends	¾ × 4	× 15	
1 floor	48- × -60- × -¾ plywood or particle board (or boards)		

Covering: shiplap boards about ¾ × 6

to project at least 1 inch at the eaves. Leave the lower edges straight, or give them a regular pattern of deckle edges. The pattern shown is distinctive (FIG. 4-7C) and you can cut it with a portable jigsaw. Cut one and use it as a pattern for marking the others. The central piece might have a simple point or might be curved to match the other decoration.

If you put strips across the sides at seat height, you can place one or more boards across as a seat and move them to stand on end if you want to use the shelter as a store. Another idea would be to make a bench with its feet arranged to come between the sides (FIG. 4-7D), then you can lift outside when you prefer the open air.

If the sun shelter is to stand on soil or grass, soak the lower parts, at least, with preservative. For a permanent position, you should place it on a concrete base. You can leave the inside untreated, but it would look best if you finished it in a light-color paint, even if you paint the outside in a dark color.

CANOPIED SHELTER

If there is driving rain or the sun is very hot, it is advantageous to extend the roof forward on your shelter. If you can extend the sides also, you can provide a wind break, as well as improve the appearance of the shelter.

The previously listed advantages are for shelters used in a garden or yard, and anyone selling or collecting tickets at an event will appreciate them. The shelter shown in FIG. 4-8 is large enough for many purposes, but it is small enough for a single-sloped roof. This type shelter is simple, but if you want a much larger shelter, it would be better to have a ridged roof, which we will describe later. This shelter is shown on a wooden base that extends all around, with extra width at the front, so you can bring a chair forward, while still benefiting from the shade of the structure. You can make a base of concrete, or you can place the shelter directly on the ground, if you are locating it only temporarily. Construction is done by bolting sections together, so you can do most of the work on the parts before erecting them on a site.

As shown, most of the framing is 2-inch-square wood and the covering is shiplap boards. You also can use plywood or other covering material. You can board the roof, but stout, exterior plywood is strong and simple. Figure 4-8 is shown with the front board decorated with a deckle edge, but if you use a plain board, it could carry announcements if you are using the shelter at an event. Nail, halve, tenon or hold framing joints, including those on the base, with sheet-metal gussets.

The sizes suggested in FIG. 4-9A will make a shelter of reasonable proportions. If you alter sizes, make sure the roof has enough slope to allow water to run off. If you get heavy snowfalls, it might be advisable to make it steeper. Do not lengthen the canopy too much if you do not anchor down the shelter, otherwise it might tend to fall forward.

Start by making a pair of sides (FIG. 4-10A). Keep three uprights square to the bottom. Bevel the sloping front pieces and nail them at the bottom (FIG. 4-10B). At the top, place a 6-inch upright part to take the front board. When you cut that corner to shape, reinforce it with another piece inside (FIG. 4-9B).

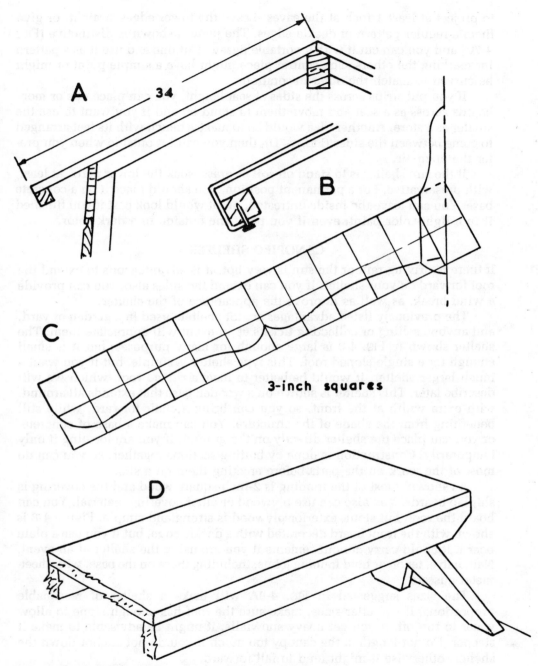

A

34

B

C

3-inch squares

D

FIG. 4-7. Roof, bargeboard, and seat details for the sun shelter.

FIG. 4-8. This shelter has its sides and roof extended to make a canopy.

Cover the framing with shiplap boards, starting from the bottom. Cut the board ends, back and front, level with the framing. On the front edges only, put on cover strips (FIG. 4-10C) with rounded, outer corners.

The upright front (FIG. 4-11A) fits between the sides. The doorway is shown with sloping sides to match the design of the shelter sides. Make the overall height to match the sides and bevel the top to match the roof slope. Arrange a crossbar to give sufficient headroom (FIG. 4-11B).

When you have assembled the shelter, bolt the front and side uprights together (FIG. 4-10D). Extend the covering boards far enough to go over the second upright, without being so long as to prevent complete tightening. When you cover the front with boards, see that the assembly is without twist.

The back is a simple, boarded frame, with strips around the outside and one central upright. Make its height to match the sides and bevel the top to suit the roof. In a similar way to the front, the boarding has to overlap a second upright

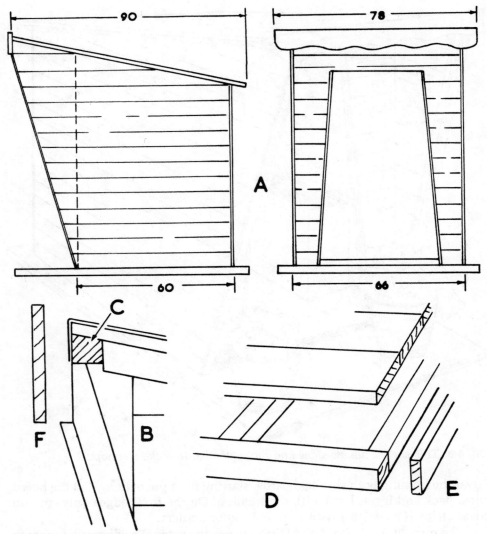

FIG. 4-9. Sizes and details of the canopied shelter.

when you bolt the parts together. Allow for this overlap (FIG. 4-10E), but you will get the neatest appearance by fitting a square piece in the corner.

When you assemble the shelter, put a bar across the front (FIG. 4-9C). This bar holds the sides at the correct distance and provides a place to attach the roof covering.

If you make a base, allow it to project about 6 inches outside the back and sides. At the front, it might go as far as the canopy. The suggested construction, has a framework with supports across at about 18 inches (FIG. 4-9D), close boarding over that, and a strip covering the board ends (FIG. 4-9E).

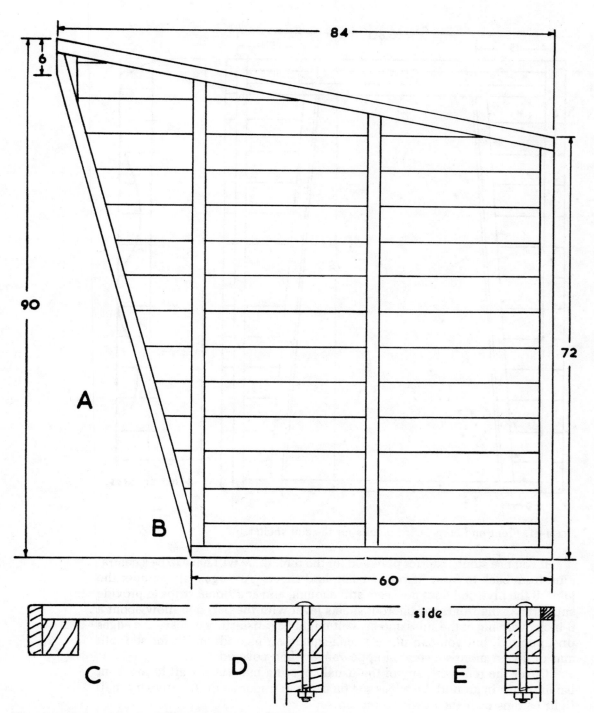

FIG. 4-10. Side and joint details of the canopied shelter.

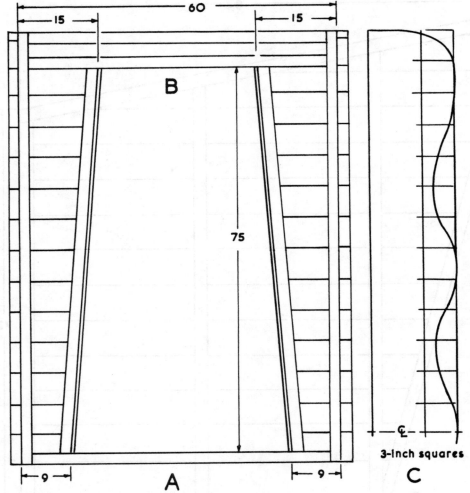

FIG. 4-11. *Front and bargeboard details for the sun shelter.*

If you use stout, exterior plywood for the roof, there will have to be a central joint from back to front, if you cut standard sheets. Arrange a strip under the joint. If the plywood does not seem stiff enough, add additional strips to provide support. At the front, cut the roof sheets level with the bar, but allow about a 6-inch overhang at the sides and back. It might be satisfactory to leave the edges unsupported, but you can place 1-inch-square strips underneath for strength and to give a more substantial appearance when covered.

Cover the roof with any of the usual covering materials. Nail to the front bar and turn in around the edges so you can nail underneath. If necessary, nail light battens over the covering on top.

Make the front board so it overhangs a little at its ends and projects above the roof (FIG. 4-9F). If you wish, shape its lower edge (FIG. 4-11C), then nail or screw it in place.

No internal work is shown, but you could build in seating or arrange removable benches, as suggested for the previous shelter.

After treating with preservative or paint, bolt or screw the shelter to its base. If you will want to move the shelter and base, use a few blocks or battens to keep the bottom edges correctly located on the base.

Materials List for Canopied Shelter

10 side frames	2	×	2 × 90	
2 side frames	2	×	2 × 62	
4 front frame	2	×	2 × 86	
3 front frames	2	×	2 × 66	
3 back frames	2	×	2 × 74	
2 back frames	2	×	2 × 66	
1 front bar	2	×	2 × 66	
2 corners	1	×	1 × 74	
2 cover strips	¾	×	3 × 90	
2 cover strips	¾	×	3 × 80	
1 front board	1	×	8 × 80	
2 roof panels	40- × -90- × -¾ or 1 plywood			
1 roof-joint cover	2	×	2 × 66	
1 roof-joint cover	2	×	2 × 30	
2 base frames	2	×	2 × 80	
6 base frames	2	×	2 × 90	
2 base covers	1	×	3 × 90	
Baseboards	1	×	6 (approximately)	
Shiplap boards	1	×	6 (approximately)	

RIDGED-CANOPIED SHELTER

A ridged roof has a more attractive appearance than a lean-to or single slope. Both look better than a horizontal roof. Over a certain size, a ridged roof is preferable, as it sheds rain and snow easier and offers less wind resistance. A shelter with a ridge from back to front and extending canopy or porch has good access and provides maximum shelter and it looks good.

The shelter in FIG. 4-12 has a floor area of about 60 inches × 84 inches, with a lengthwise canopy of approximately another 24 inches. Good clearance is provided through the doorway and it has ample headroom inside. No base or floor is shown, but you could add a wooden or concrete platform or you might build in a wooden floor. Construction is sectional, so you can prefabricate most parts and bolt them together on-site.

Shiplap boarding is suggested for the covering, but you can use other materials. The sizes given in FIG. 4-13A allow for economical cutting of standard plywood sheets. Arrange internal framing of some panels to suit joints between plywood sheets. Join framing parts to each other by halving, tenoning,

FIG. 4-12. *This canopied shelter has a ridged roof.*

or by using sheet-metal gussets. Many parts will be satisfactory if you notch and nail them.

Start by making the front (FIG. 4-14A). Fit this piece between the sides as previously done in FIG. 4-10D. Allow the covering boards to extend to overlap the side uprights. The doorway is shown 48 inches wide, but you can alter that. If you want good protection for articles stored inside, make it narrower. If you want to let in plenty of sunlight while you sit inside, make it wider.

At the top, cut away for a 2-inch × 4-inch ridge piece to pass through and put a supporting piece across, below the gap (FIG. 4-14B). After covering with boards, put rounded-edge pieces at each side of the doorway (FIG. 4-14C).

The back has the same outline as the front. Bolt its uprights to the uprights at the rear of the sides, and cut the boarding to allow for fitting a square strip in the corner (FIG. 4-10E). To allow for fully boarding the back, arrange two uprights to the full height intermediately (FIG. 4-14D).

Make the pair of sides (FIG. 4-15A), cutting the boards level with the framing. Allow a 4-inch vertical part at the front to take the bargeboards. Reinforce with a block inside (FIG. 4-15B). Check the side heights against the matching parts of the front and back. Bevel the top edges to suit the slope of the roof. Cover the front sloping edges for a neat finish (FIG. 4-15C).

For assembly, it should be satisfactory to use ⅜-inch-coach bolts at about 18-inch intervals. Drill holes for these bolts in the sides. Do not continue drill-

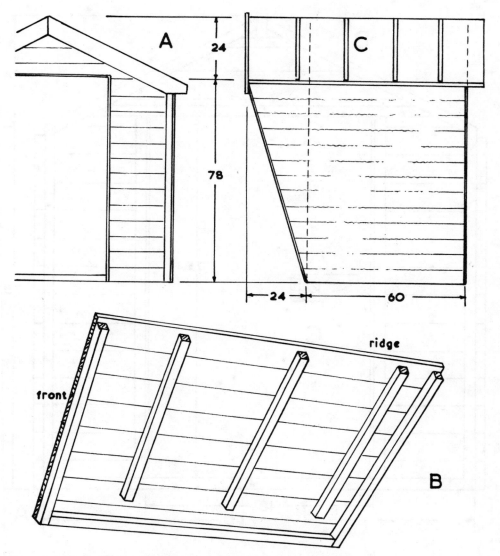

FIG. 4-13. Sizes and roof details for the ridged-canopied shelter.

ing into the front and back uprights until you bring them together on-site, to ensure exact mating of holes.

Make a 2-inch × 4-inch ridge piece to go right through the full length of the roof. Bevel its top edge to match the slope on each side. At the front, match the extension with the shelter sides. At the back, an overhang of about 5 inches should be enough.

You can make the two halves of the roof of thick plywood or wide boards (FIG. 4-13B). Bevel the top edges so they meet along the ridge. Make the sections wide enough to allow for a 5-inch overhang at the sides. Frame at back, front,

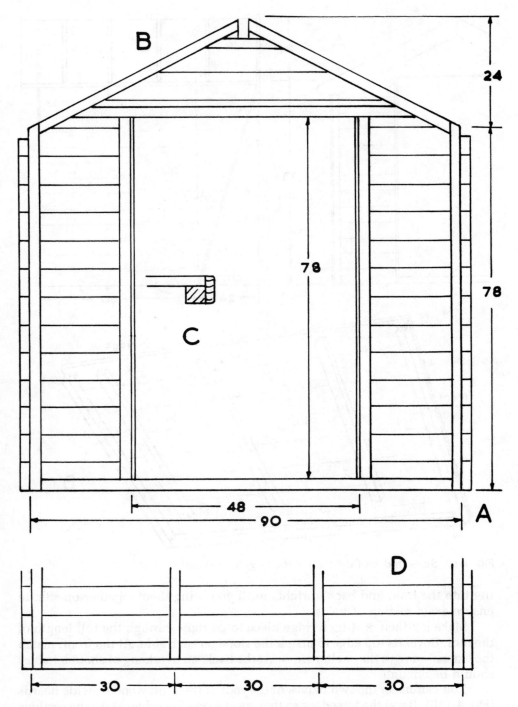

FIG. 4-14. Front and back details for the ridged-canopied shelter.

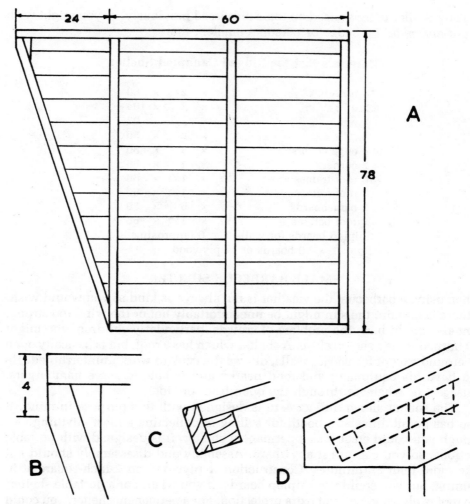

FIG. 4-15. *Side and roof details for the ridged-canopied shelter.*

and eaves. Arrange a strip to fit inside the shelter front and another to fit inside the back. These strips will hold the whole assembly square. The number of other pieces will depend on the stiffness of the roof, but there will have to be at least one more.

Cover the roof from eaves to eaves, with a good overlap. Turn under and fix with large-headed nails at the eaves and the ends. To hold down the fairly large area on top, nail light battens at about 18-inch intervals (FIG. 4-13C).

Fit bargeboards at the front only, or at both ends. Cut so the boards stand about 1 inch above the roof surface and overlap about 1 inch at the corners of the roof (FIG. 4-15D). Nail into the roof end and into the shelter's sides. No central, decorative piece is shown at the apex of the bargeboards, but you can use one similar to those used on roofs discussed previously; or you might wish to cut and mount a badge or personal emblem there.

Materials List for Ridged Canopied Shelter

14 uprights	2	×	2	×	80
2 uprights	2	×	2	×	104
5 rails	2	×	2	×	92
4 rails	2	×	2	×	56
4 edge covers	1	×	3	×	80
2 corners	1	×	1	×	80
10 roof frames	1½	×	1½	×	60
2 roof frames	1½	×	1½	×	90
2 bargeboards	1	×	6	×	60
8 roof battens	½	×	1½	×	60

Shiplap boards for walls 1 × 6 (approximately)
Roof: 1- × -10 boards or ¾ plywood

SMALL BARBECUE SHELTER

When using a barbecue, the weather is not always as kind as you would wish. If there is no rain, the sun might be uncomfortably hot or the wind too strong. You also might have the problem of crowds, particularly children who might get burned if they get too close. A shelter which has a roof, but is basically open all around except for barrier walls, allows the cook to work uninterrupted. He can have his equipment and food nearby and is able to serve hamburgers, sausages, or whatever through the open front or sides.

The shelter shown in FIG. 4-16 is designed with this purpose in mind. It also has possibilities as a booth for selling or collecting almost anything. Although you could erect it as a permanent shelter, it is designed with portable sections, so you can use it anywhere. Assembly and disassembly should not take more than 30 minutes. Construction is plywood on 2-inch-square strip framing, but you could use shiplap boards, if you wish. For a portable shelter, the roof is plywood without extra protection. For a permanent shelter, you could stiffen its edges and cover the roof with roofing felt or other material.

The suggested sizes are based on 30-inch-wide bays (FIG. 4-17A). The height allows normal clearance through the doorway at the back. The barrier-wall edges are 36 inches from the ground, which is a convenient height for serving and probably enough height to keep most wind off the barbecue surface. Boarding directly above the barrier walls is not included, but you could fully enclose any of the walls, if you wish, particularly if you have to contend with a strong, prevailing wind in one direction. You could arrange removable panels to fit in or hang on the open framing, if your needs will vary.

Start with a pair of ends (FIG. 4-17B and 4-18A). With the complete structure, make sure the parts of the uprights that edge the openings are smooth and with lightly-rounded edges, so they are safe if anyone holds on to them. Halve together the uprights and rails which cross at the middle. You can make

FIG. 4-16. This barbecue shelter has a roof and barrier walls, but there is plenty of open space for ventilation.

any of the other usual joints elsewhere. Nailing on the plywood skin will give rigidity to the assemblies.

Hold corner joints between walls with ⅜-inch bolts at about 18-inch intervals, similar to the other shelters described earlier. For a portable shelter, you might finish the covering panels level with the edges. For a more permanent shelter, the plywood on the ends might overlap the back and front for a neater appearance. If you plan to keep the shelter in sections for transport by truck to anywhere you need it, you might damage the overhanging plywood, so it is better to omit it.

The back is a simple frame, mostly open (FIG. 4-18B), so use strong joints, particularly at the top. The lower plywood panels should prevent distortion of the assembly.

The front (FIG. 4-17C and 4-18C) has plywood top and bottom to give rigidity. Check the back and front heights against the ends, and bevel the top edges to match the roof slope. It will be best to locate bolt holes during a trial assembly.

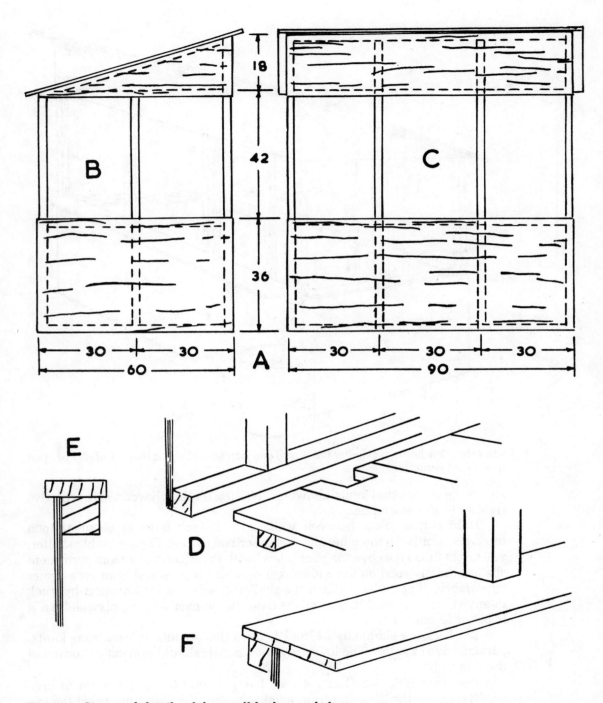

FIG. 4-17. Sizes and details of the small barbecue shelter.

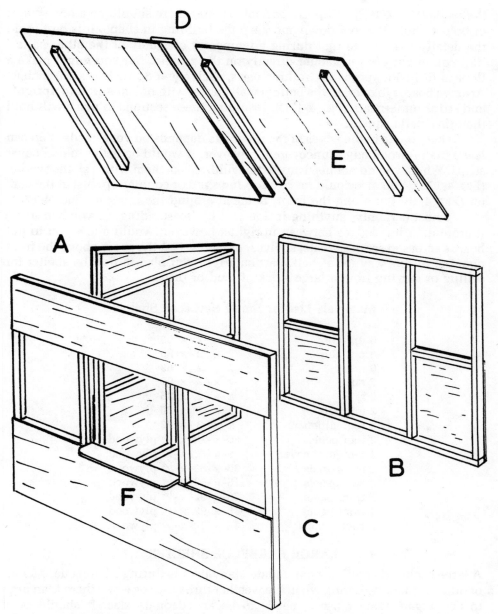

FIG. 4-18. *Assembly arrangements of the small barbecue shelter.*

The roof can be 72 inches down the slope and in two pieces to overhang about 6 inches at the ends. Put a 6-inch-wide cover strip on one half to overlap the other piece (FIG. 4-18D). A stiffening strip under the lapped piece will prevent distortion. If this will be a permanent shelter, you can screw the roof into place as it is, but if it is portable, put more strips underneath to fit inside the walls (FIG. 4-18E). With the portable shelter there should be a few screws or bolts to hold the roof down and keep the building in shape. It is best to do the detail work on the roof during a temporary assembly of the other parts.

You might use the shelter directly on the ground, but you might make a floor to fit inside, probably in three parts, each about 30 inches × 60 inches. Arrange boards to overlap the bottom wall edges, with notches around uprights and strips underneath (FIG. 4-17D). Dropping these sections in place will hold the other parts square.

What you do at the edges of the openings depends on your needs. You can leave the plywood edges uncovered, however, it would be better to add cover strips. Where you do not need any extra width, make them with a slight overlap (FIG. 4-17E). You also could make a serving shelf or countertop, just at the center (FIG. 4-18F) or along the front, notched around the uprights (FIG. 4-17F).

Almost certainly, anything inside will be freestanding, so you can move it around. A few ledges between uprights, however, would allow you to put boards across to serve as seats, shelves, or tables. Use the space above the front opening in this way for a shelf, particularly if you plan to use the shelter for selling or serving from a large stock of food or drink.

Materials List for Small Barbecue Shelter

6 upright	2 × 2 × 98
8 uprights	2 × 2 × 80
6 rails	2 × 2 × 62
7 rails	2 × 2 × 92
2 rails	2 × 2 × 36
2 top rails	2 × 2 × 70
3 roof stiffeners	2 × 2 × 62
2 roof panels	50- × -72- × -½ plywood
1 roof-joint cover	6- × -72- × -½ plywood
2 end panels	36- × -62- × -½ plywood
2 end panels	18- × -62- × -½ plywood
1 front panel	36- × -92- × -½ plywood
1 front panel	18- × -92- × -½ plywood
2 back panels	32- × -36- × -½ plywood

LARGE BARBECUE SHELTER

A large shelter that will accommodate many people during a barbecue, picnic, or other outdoor gathering when the weather turns sour or when there is an urge to congregate more closely, needs to be an adequate size. It should be a permanent structure and should be built in position, although you might prefabricate some parts.

The shelter shown in FIG. 4-19 has a ridge roof and a doorway at one side, although you can place these items elsewhere without difficulty. Closed walls reach 42 inches high and there are open spaces above that. Spacing allows for seating to be built in, or you might prefer to use separate chairs or benches. The framework has to be stouter than in previous shelters and is mostly 2 inches

FIG. 4-19. *The large barbecue shelter has space inside for several people as well as the barbecue.*

× 3 inches. The cover might be shiplap boards, although you can use plywood or other covering. The sizes suggested in FIG. 4-20 would allow you to use a single, standard plywood sheet in the width and one and a half sheets in the length. The roof is boarded and covered with bitumastic felt or other roofing material. The pair of gables settle the shape, but two trusses to support the ridge and purlins are under the roof. You should install this type of building on a concrete base. A wooden floor would be inappropriate. In a suitable situation, compacted earth might be satisfactory.

Start with the two gables, which are the same (FIG. 4-21A). All of the framing is 2-inch × 3-inch strips, except the piece across below the eaves, which is a 2-inch × 6-inch section. All pieces have their 2-inch faces towards the outside. You might halve corners (FIG. 4-22A). Halve the central upright where it crosses the rails. At the top, put cheek pieces so you have a space for the ridge (FIG. 4-21B). Notch the other uprights and nail them to the rails. Notch the ends of the 6-inch piece (FIG. 4-22B). Check squareness of the framework and see that the opposite ends match as a pair. Board the outsides level with the openings and edges.

The lengthwise supports for the roof are the ridge, which fits into the apex slots of the gables, two wall plates (FIG. 4-22C), which rest on the side piece

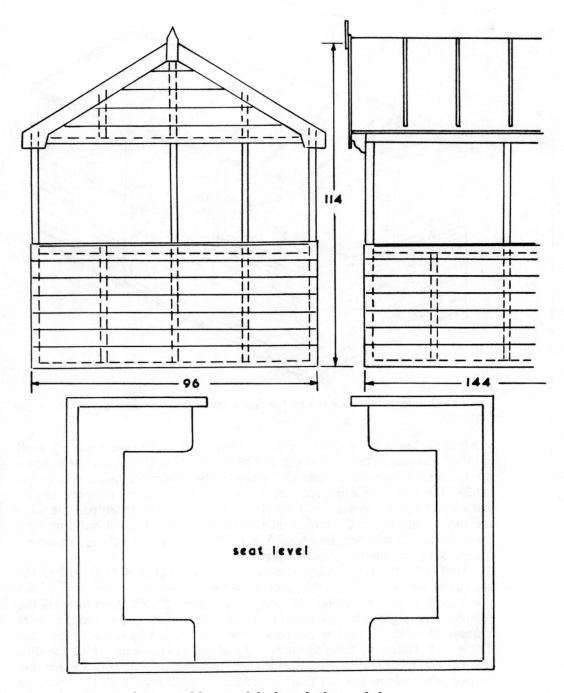

FIG. 4-20. Sizes and suggested layout of the large barbecue shelter.

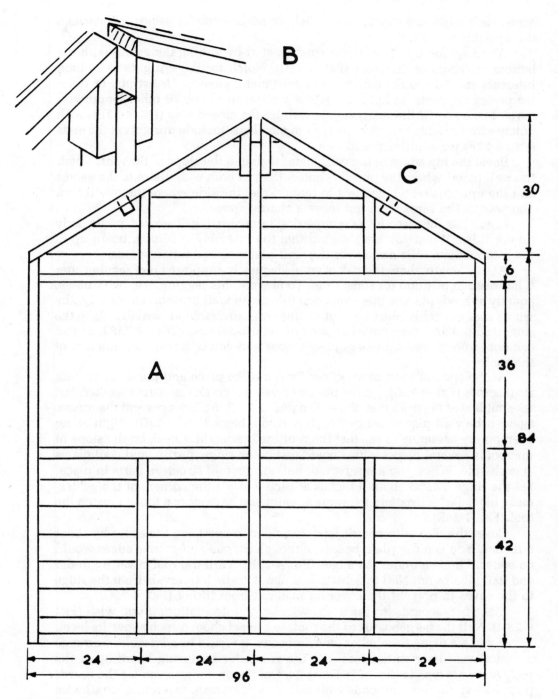

FIG. 4-21. An end of the large barbecue shelter.

across each gable, and two purlins. Notch the purlins into the gables below where the uprights come (FIG. 4-21C)

When setting up the building on-site, start by joining the gables with the bottom members of the sides (FIG. 4-22D). Notch for the uprights at 24-inch intervals and join to the gables with sheet-metal gussets. Use struts, or other temporary supports, to hold the gables vertical until you fit other lengthwise parts. Make and fit the intermediate rails on the closed side (FIG. 4-22E), with notches for uprights to match the bottom members. Include uprights at the ends (FIG. 4-22F) for additional stiffness.

Bevel the top edges of the wall plates to match the slope of the roof. Notch the wall plates where the uprights come, then fit both wall plates to the gables and the uprights which connect to them. Treat the side where there will be a doorway in the same way, but leave a central space.

Check that the assembly is symmetrical by comparing diagonals, particularly on the sides. When you are satisfied that the assembly is square, board up to the intermediate rails and cover the wall plates with similar boards.

At this length, there is a risk of an inadequately-supported roof sagging after it has been in position for some time. To prevent this sagging, use two trusses, equally spaced, placing one over each full-depth wall upright. Obviously, the whole roof assembly must be kept in line. It is advisable to partially make the ridge and purlins, then have the parts of the trusses ready (FIG. 4-23A), so you can cut joints by testing where pieces cross and you will maintain the line of the roof.

At the apex of each truss, allow for the ridge piece and put a supporting strip across (FIG. 4-23B). At the purlin crossings, notch the parts together, but take out less of the truss than the purlin (FIG. 4-23C). At the eaves, cut the trusses against the wall plates and nail each part to a block (FIG. 4-23D). Sight along a temporary assembly to see that the roofline is straight. Check on the slope of the roof in several places with a board resting on ridge, purlin, and wall plates at each side. When you are satisfied, nail all the roof structure parts in place. Let the ridge extend about 5 inches at each end. At the corners of the gables, make blocks which extend the same amount and have sloped tops to match the roof (FIG. 4-23E).

Cover the roof with boards from the ridge to overhang the eaves by about 6 inches. You can use plain boards, although tongue-and-groove edges would be better. Thicken under the edges (FIG. 4-23F). Turn the roof covering under and nail all around. Nail thin battens at about 18-inch intervals from the ridge to the eaves to prevent the covering material from lifting (FIG. 4-23G).

Make bargeboards for the ends, with central decoration, if you wish (FIG. 4-23H). Nail the bargeboards to the roof, to project about 1 inch above its level.

If you are using the shelter for a barbecue, it would be advisable to arrange a central ventilator in the roof. Fit the ventilator in during boarding of the roof. A simple type is shown in FIG. 4-24. Cut the opening as you lay the boards (FIG. 4-24A). Support cut ends with battens underneath, extending lengthwise on each side of the shortened boards (FIG. 4-24B). Nail on strips at each end of the opening to raise the ventilator roof about 5 inches, then board lengthwise

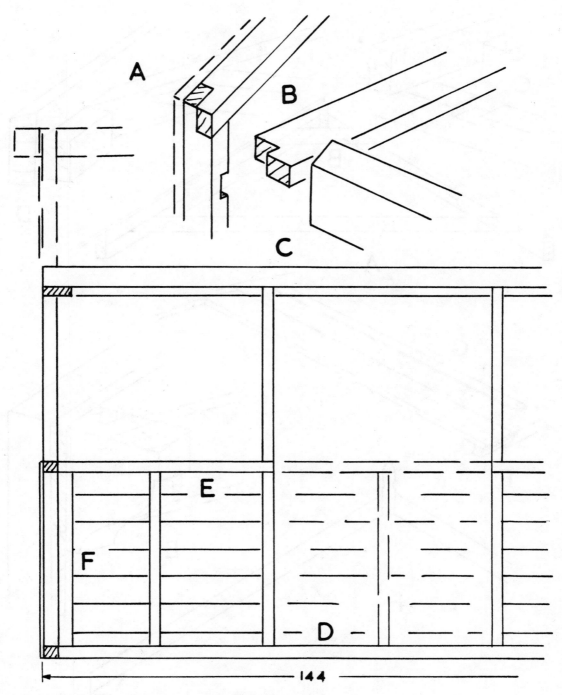

FIG. 4-22. Side arrangements of the large barbecue shelter.

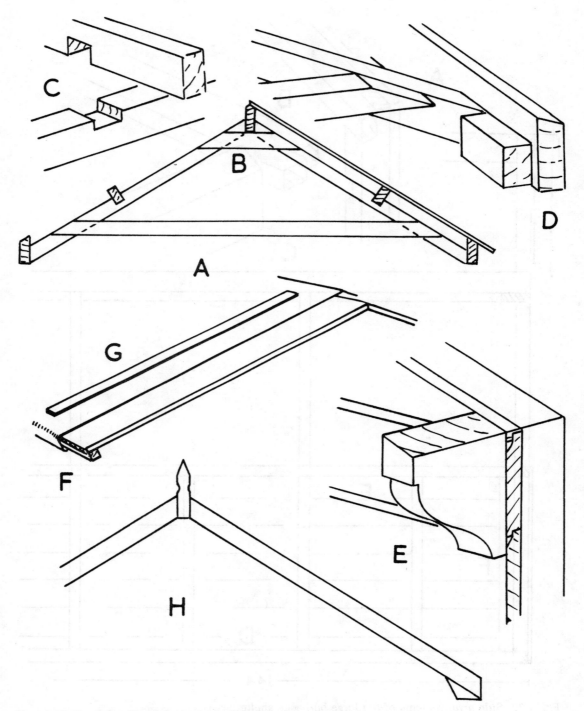

FIG. 4-23. Roof details of the canopied shelter.

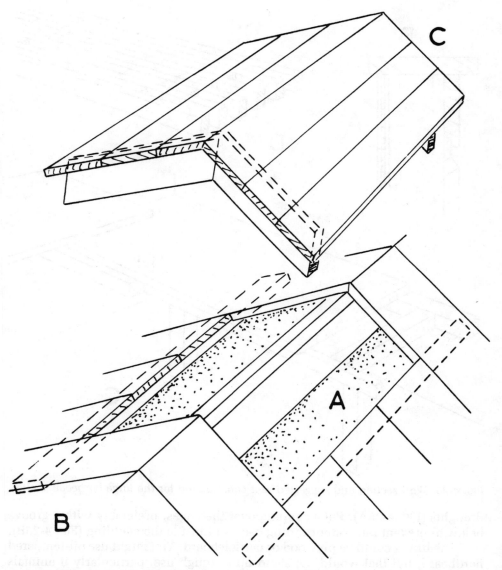

FIG. 4-24. Ventilator details of the large barbecue shelter.

over them (FIG. 4-24C). Do not cut or alter the ridge piece below the opening. When you cover the roof, carry the covering material up the ventilator and put more over its roof. The ventilator does not have to be central; you can position it to suit where you will put the barbecue fire.

What you do inside the shelter depends on its intended uses. You can leave it as it is, which might be satisfactory for varied and occasional uses. A lining inside the closed walls would improve appearance, add to strength, and reduce

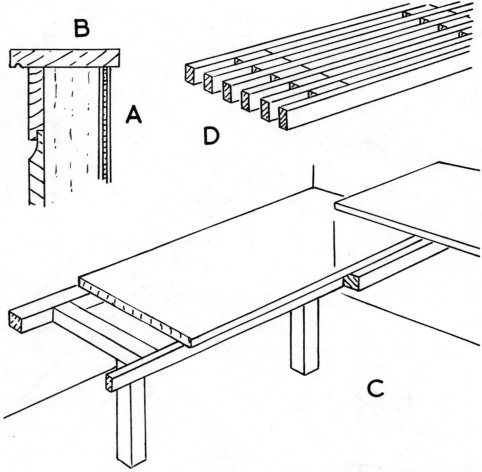

FIG. 4-25. *Wall section and suggested seat construction for the large barbecue shelter.*

draughts (FIG. 4-25A). Put a capping over the edges, preferably with a groove below to prevent rainwater running back and inside the cladding (FIG. 4-25B).

The lining could be plywood or particleboard. You might use oil-tempered hardboard, but that would not stand up to rough use, particularly if animals will use the shelter.

Bench seats are easy to build in. Put battens where the seats are to come. A seat height of 16 inches is suitable. Make framed supports. Put a strong support underneath if you carry the seating around a corner (FIG. 4-25C). The simplest seat top is made from thick, exterior-grade plywood. Boards can make up the width. A neat top, which does not trap water or leaves that blow in, is made with 1-inch × 2-inch strips on edge (FIG. 4-25D).

Arrange seats as lockers for storage. Complete them as boxes, nailing down with a few inches at the back of the top and arranging the rest to hinge up. Fit shelves inside the roof at the ends or along each wall plate.

Materials List for Large Barbecue Shelter

Gables

4 uprights	2	× 3	×	86
4 uprights	2	× 3	×	120
4 rails	2	× 3	×	98
2 rails	2	× 6	×	98
4 uprights	2	× 3	×	44
4 uprights	2	× 3	×	24

Sides

3 rails	2	× 3	×	146
2 rails	2	× 3	×	50
2 wall plates	2	× 6	×	146
8 uprights	2	× 3	×	86
6 uprights	2	× 3	×	44
4 blocks	2	× 6	×	12

Roof

1 ridge	2	× 6	×	158
2 purlins	2	× 4	×	158
4 truss rafters	2	× 3	×	60
2 truss ties	2	× 3	×	86
46 roof boards	1	× 6	×	64
4 roof edges	1½	× 1½	×	64
2 roof edges	1½	× 1½	×	158
16 roof battens	½	× 1	×	64
4 bargeboards	1	× 6	×	68

Walls

Shiplap boards 1- × -6 or ¾ plywood (approximately)

Ventilator

4 ends	1	× 5	×	24
2 battens	1	× 2	×	30
6 tops	1	× 6	×	24

5
WORKSHOPS

A COMMON SMALL BUILDING NEED IS A WORKSHOP. A WORKSHOP MIGHT BE FOR ANY craftwork, such as woodwork or metalwork, but an enthusiast for macramè, leatherwork, or collecting and repairing all sorts of things might want a shop where he can carry on his hobby. Whatever the craft or activity, the advantage of a separate workshop is that you can equip it as you wish, and then leave it ready for use the next time. You do not have to clear things away, as is necessary in a room you need for other purposes as well.

If yours is a dirty or noisy activity, the separate shop avoids annoying others in the house. If you need to keep bulky stock, as you might for woodworking, storage in the shop is tidy and away from other people. If you use machines, the separate building gives you scope to arrange them permanently so you can use them to their best advantage. The considerable noise you often create will be kept away from those who do not appreciate it, much more so than using a room in the house or basement.

Whether your craftwork is a hobby or a part-time or full-time business, consider the space you need and your probable future needs. Most experienced craftwork enthusiasts soon reach a stage where they wish they had more space. If it is a woodworking shop, consider any machines you have now and what you might get. A guiding size is a 48-inch-x-96-inch sheet of plywood or manufactured board. Ideally, you should be able to move it in any direction around a table saw. Practical considerations might dictate a building too small to permit this turning, but you should think about how you will deal with large and long pieces of wood. Allow for assembly space. Do not fill the shop with equipment so that you cannot put together a table or cabinet. If the weather is reasonable, you might have to manipulate large things through doors or windows.

Obviously, a shop should be strong and weatherproof, but besides standing up to anything that might happen outside, some activities might require strength inside. Scrap wood or metal accidentally flying off a machine might hit a wall or roof with considerable force. Hardboard is unlikely to make a strong enough lining for your workshop. Strength in the building also is valuable when you want to brace something to it. If you lever against a wall, you do not want the wall to distort.

There should be plenty of light, both natural and artificial. Windows normally should be above bench and working level, to reduce the risk of them becoming broken too often. You do not want the shop to be a greenhouse, but make sure enough light gets in. It could be dangerous working in shadows. Windows in the roof can provide plenty of light, with a good spread, but they are not so easy to make watertight. In most shops, have some opening windows. Besides ventilation, you might need them to get long materials in and out of the shop.

If you live where there is a mild climate, a building with little or no insulation might be all you need, providing it is waterproof. Elsewhere, you should consider insulation, particularly if you suffer from extremes of climate and you want to be able to use the shop all year round. Windows are a source of heat loss, so do not make them any bigger than necessary. Double-glazing is possible, but unusual in this type of building. Curtains or sheets of hardboard over the glass will have a similar insulating effect when you are working by artificial light. Remember roof and floor insulation.

Although a concrete base is desirable as a firm foundation for a shop, its unprotected surface inside the building is undesirable. If you drop tools and other items on it, you might damage them. The surface is cold and uncomfortable to your feet. It might produce dust that you do not want in some activities. It would be better to have a wooden floor, preferably with an air gap over the concrete. If you cannot arrange that, provide some protection with boards or plywood directly on the concrete. Rubber mats will prevent much damage to dropped chisels or other edge tools.

You will need electricity in the building. Make sure you install this properly and adequately for your needs. A temporary cable from the house supply could be dangerous and almost certainly would not cope with all your needs. Have plenty of lamps. Individual lamps on adjustable arms are better than a few fixed, general lights. Do not rely too much on fluorescent lighting, particularly if it is close to machines because of the risk of the *stroboscopic effect*. You could meet a situation where light and machine frequency agree and you assume a part is not moving, when it is. Individual filament lamps might be better. Install the necessary main switches, trip switches, fuses, etc. for more than your initial needs, so there will be no problem when you add more load, as you certainly will.

It is easy to be so occupied with all the practical needs in your new workshop, that its effect and appearance on others not so enthusiastic on functional aspects might be lost. Locate the building and arrange its external appearance so it is as attractive as possible.

BASIC WORKSHOP

The size and arrangement of a building you make for use as a hobby shop will depend on many factors, including the available space and the situation. You will have to consider the actual craft or occupation and its needs. However, a building with about an 8-feet × 12-feet floor area with working headroom and several windows, will suit woodworking and metalworking as well as many other crafts. The building shown in FIG. 5-1 is of basic, partly prefabricated construction. It has a door wide enough to pass most pieces of furniture or light machinery and the suggested windows should give enough light if most activities are on a bench which you arrange at one long side.

At the entrance end, a window is shown in the door. The wall alongside it then would be available for shelves and racks. Two windows which open are shown over the long bench (FIG. 5-2A) and you might put another window which opens at the back (FIG. 5-2B). Besides providing ventilation, these windows allow long or awkward work to be extended outside, if that is the only way to handle it. You might leave the other long side without windows, but that will depend on your needs. If you want to have a lathe or table saw near that wall, arrange more windows there, not necessarily ones which open.

This structure is not intended to be a portable building. It is not intended to be moved once you have assembled it fully. However, you can prefabricate much of it. You can make the four walls elsewhere, then assemble them to each other on-site and add the roof. Nearly all the framing is made from 2-inch × 3-inch-section wood. The covering is shiplap boards about 6 inches wide, but

FIG. 5-1. A basic workshop with boarded walls and ample windows.

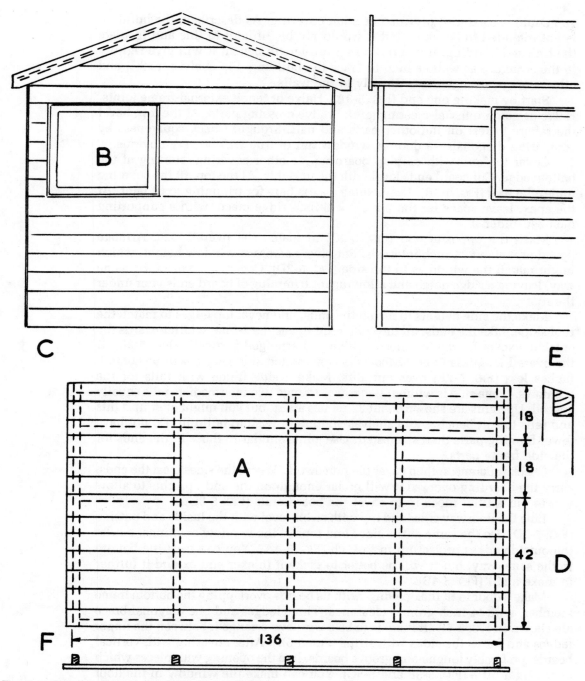

FIG. 5-2. Suggested sizes for the basic workshop.

you could use exterior plywood or other covering. As described, the building is not intended to be lined, but it would not be difficult to line and insulate the finished building. If you build in a full-length bench, it will give rigidity to the structure as well as help to brace the building. Fix shelves, racks, and other storage arrangements directly to the walls.

Start by making one end (FIG. 5-3A). Halve or tenon external-frame joints. Halve or notch internal-meeting joints. Halve crossing parts. At the top, bevel the rafters to rest on the other parts and nail through. Check squareness by comparing diagonals—a door or window out of true will be very obvious.

Cover the end with shiplap boarding, or other covering, starting at the bottom edge. Cut board ends level with the uprights. At the top, fit the covering under the roof (FIG. 5-3B). Leave some excess here for trimming to fit later. At the apex, leave space for the 2-inch × 6-inch ridge piece, with a supporting member under it.

Make the opposite end (FIG. 5-2C) to match the overall size. Arrange uprights at about 24-inch intervals. Put pieces across at window height, which might match the windows in the side (FIG. 5-2D). Cover this end in the same way, leaving a ridge notch and allowing for trimming of board ends later under the roof.

Make the side heights to match the ends, and bevel top edges to match the roof slope. Like the ends, all the side framing has the 2-inch width towards the outside, except for the top piece, which you arranged vertically (FIG. 5-2E). If the overall length is to be 12 feet, the constructed side length will be about 8 inches less (FIG. 5-2F) over uprights. Make a side frame with rails for the windows. If one side is without windows, arrange two intermediate rails equally spaced. Uprights are shown about 32 inches apart, but you might alter uprights and rails to suit benches and shelves you might wish to build in. Do not have fewer framing parts than suggested. Use joints similar to those in the ends for the side frame parts.

Check squareness, then cover the framework. Where the sides meet the ends, carry the boarding over, so it will go far enough on the end uprights to allow you to put a filler piece in to cover the board ends (FIG. 5-3C).

Line the doorway sides and top with strips level with the inside and outside (FIG. 5-3D). Do the same at the sides and tops of the window openings, but let the outside edges project up to ½ inch (FIG. 5-4A). You can treat the bottom in the same way, but it will be better to make it thicker and extend it further to make a sill (FIG. 5-4B).

Make a door to fit the opening, with its boards overlapping the bottom frame member, with ½-inch ground clearance. Three ledges and one diagonal brace are shown in FIG. 5-5A. If there is to be a window, arrange it between the upper ledges and frame the sides with strips (FIG. 5-5B). After covering with vertical boards, preferably tongue-and-groove boards, line the opening with pieces which overhang a little (FIG. 5-3E and 5-5C). You can make the window in the door by simply holding glass between strips (FIG. 5-3F). Cut the glass a little undersize, to reduce any risk of cracking. Waterproof the window by embedding the edges in putty or a jointing compound.

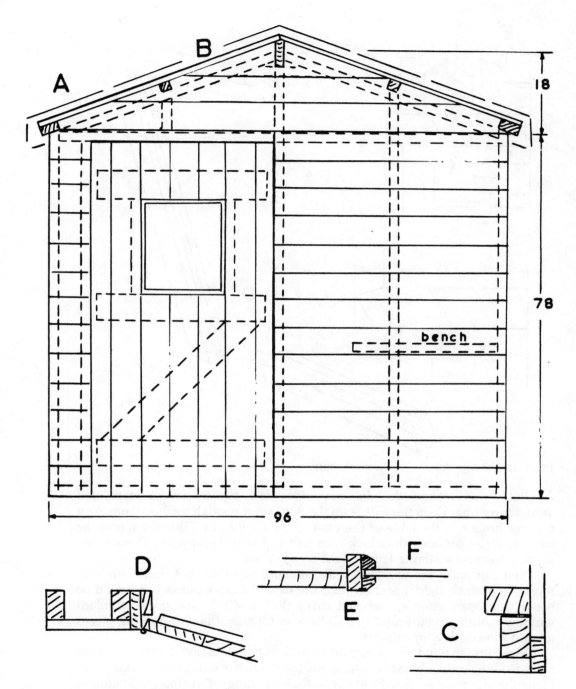

FIG. 5-3. *The door end of the workshop.*

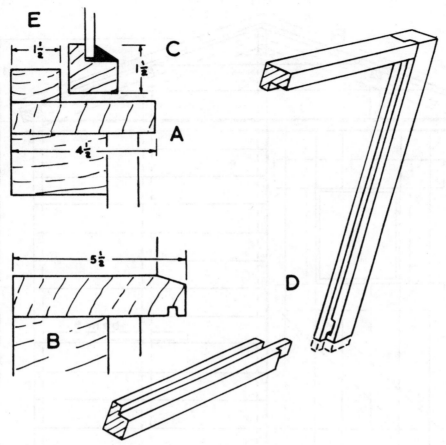

FIG. 5-4. *Window details for the workshop.*

Put strips around the doorway sides and top to act as stops and draughtproofing. Keep the ledges on the door short enough to clear them. You can put hinges in the edge of the door, or you might fit T hinges across the surface. Fit an ordinary door lock with bolt and key, if you wish to secure the shop, otherwise a simple latch might be adequate.

You can make the windows with standard molding, but these windows might be a much lighter section than the usual house windows. It would be better to prepare simple, rabbeted strips (FIG. 5-4C). If you use a standard window molding, you probably will have to increase the width of the pieces around the window openings.

Make up the windows with mortise-and-tenon joints (FIG. 5-4D). Leave the sides too long until after assembly, to reduce the risk of end grain breaking out. Make the windows so they fit easily in their openings. Put stop strips around the inner edges of the framing (FIG. 5-4E). Hinge the windows at the top and arrange fasteners and struts inside at the bottom. You might want to lift the windows horizontal occasionally, but you can do that with a temporary strut

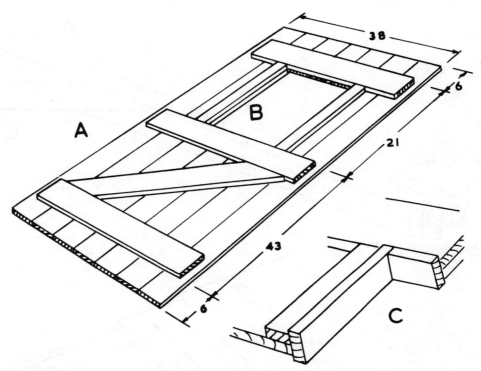

FIG. 5-5. *The workshop door with a cutout for a window.*

or a cord from higher on the wall. When you are satisfied with the fit and action of a window, you can putty in the glass, although it might be better to putty after you have painted the wood. The building will look attractive if you paint the window frames and bargeboards a different color than the main parts, so you could paint the window frames and glaze them in advance of final assembly.

The roof is supported by the 2-inch × 6-inch ridge, 2-inch × 4-inch eaves laid flat, and 2-inch-square purlins halfway down each side of the roof. Nail the eaves strips and purlins to the sloping top frames of the ends and bevel the ridge to match (FIG. 5-6A). Let the ends project about 3 inches at each end of the building. Cut the shiplap-covering boards around them and trim their top edges to match the roof (FIG. 5-6B).

On a 12 foot length, it should be sufficient to prevent sagging of the roof by having rafters only at the center. If you make the building longer or have doubts about the stiffness of the assembly, use two sets of rafters, spacing them equally. Cut a pair of rafters to fit between the top pieces of the side frames and the ridge (FIG. 5-6C). Check straightness of the sides while cutting. If you get the length of a rafter wrong, it could make the side bulge or bend in slightly. You can have a nailing block at one or both ends of each rafter. A block below the purlin will locate and support it (FIG. 5-6D). Put a strip across the rafters below the ridge (FIG. 5-6E). No other lower tie-down is needed.

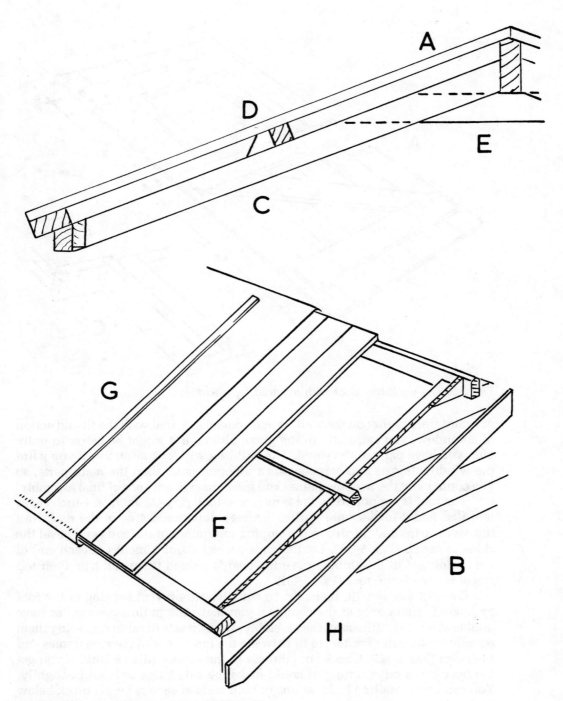

FIG. 5-6. *Roof details for the workshop.*

Materials List for Basic Workshop

Ends

5 uprights	2	× 3	×	80
2 uprights	2	× 3	×	100
3 uprights	2	× 3	×	90
4 rails	2	× 3	×	98
4 tops	2	× 3	×	54

Sides

10 uprights	2	× 3	×	80
8 rails	2	× 3	×	138
4 corners	1	× 2	×	80

Cladding

1- × -6 shiplap boards or ¾ plywood

Door

2 frames	1	× 4	×	80
1 frame	1	× 4	×	40
3 ledges	1	× 6	×	38
1 brace	1	× 6	×	45
7 boards	1	× 6	×	80
2 window sides	1	× 2	×	16
4 window frames	1	× 2½	×	16
8 window fillets	1	× 1	×	16

Windows

6 frames	1	× 4½	×	20
3 frames	1	× 4½	×	36
1 sill	1	× 5½	×	36
1 sill	1	× 5½	×	74
6 stops	1	× 1½	×	20
6 stops	1	× 1½	×	36
6 window sides	1½	× 1½	×	20
6 window rails	1½	× 1½	×	36

Roof

1 ridge	2	× 6	×	160
2 eaves	2	× 4	×	160
2 purlins	2	× 2	×	160
16 battens	½	× 1	×	54
2 rafters	2	× 3	×	54
1 tie	2	× 3	×	40
4 bargeboards	1	× 5	×	60
Covering	1- × -6 boards or ¾ plywood			

You can cover the roof with exterior-grade plywood, but FIG. 5-6F shows it boarded. Finish level at the eaves. Put covering material over the structure in single lengths from one eaves to the other, if possible, turning the ends under and nailing them. Any overlaps should be wide and you should arrange them so water cannot run under. Nail on battens (FIG. 5-6G) at about 18-inch intervals to prevent the covering material from lifting.

Simple, narrow bargeboards are suggested in FIG. 5-6H, nailed to the roof ends after covering. You could make more elaborate ones, as shown on some earlier projects to use your own ideas. Make sure there is clearance for the door to swing open, at least to 90°.

LARGE-DOOR WORKSHOP

Some activities involve the movement of large items in and out of the shop. You might wish to work on a large piece of machinery or a car conversion. You might want to build a boat. Even on something like a luggage trailer, prefabricated parts of another building, or parts of a deck, you need access doors larger than the ones you would use for more modestly-sized projects.

The shop needs large doors, but if the project will be inside for a long period, you might not want to use the large doors for normal access, so you need a smaller door elsewhere. If the project and all that goes with it will occupy a lot of space, you will find it advantageous to have your bench and most machines at the back of the shop. The second door allows you to get in and out away from the clutter in the main area of the building.

It is possible to arrange the layout of a ridged or lean-to building to suit this sort of work, but the design shown in FIG. 5-7 has a flat, moderately-sloped roof and a porch. This design allows the double doors to be big enough, yet you can keep them shut to protect you against bad weather. The porch lets you get in and out away from the bench and machines. It might shelter you from rain and wind, and you could hang clothes on it.

Size will depend on your needs and the available space, but those shown in FIG. 5-8A are suggestions which you can modify considerably. The high end will have to be big enough to allow the doors to admit whatever you are making or working on. At the other end, you must have adequate headroom. Couple these requirements with the slope of the roof. You need a moderate slope, even if your rainfall is slight and infrequent. Steeper slopes are advisable for heavy rain areas. Snow will not slide off this type of roof; you must make it strong enough to take the weight. Whatever your climatic situation, it is unwise to have a horizontal roof.

The design shows a shop area about 7 feet × 10 feet, with a porch 3-feet square (FIG. 5-8A). It is assumed that the building will be clad with ¾-inch-exterior plywood, although you could use shiplap boarding or other covering. The framing is mostly 2 inches × 3 inches, arranged with the 3-inch side in the front to back direction of the building. The sides have the 3-inch side towards the skin, and the front and back assemblies have the 2-inch side towards the skin. The bottom rails of all parts, however, have the 3-inch side towards

FIG. 5-7. *This workshop has large doors, a sloping roof, and a porch entrance at the back.*

the foundation. The main doors are about 66 inches wide and 7 feet high. The door under the porch is 27 inches wide. Two windows are at each side and two or three are at the back. You can arrange windows to suit your needs by adjusting framing when making the walls.

You might prefabricate the sides and back and front, then assemble them on-site and add the porch and roof. The pair of sides are the key assemblies which govern the shapes of other parts, and you should make them first.

Make the side that will have the door under the porch (FIG. 5-9). Use any of the usual joints between the frame parts. Cut the sloping top to fit against the horizontal rail at the low end and nail it in place. Cover with plywood to the edges on the uprights, but at the top, the plywood should reach as high as the 3-inch rafters which you will lay on. Allow extra plywood for this area, so you can trim the edge level during assembly.

Make the opposite side to match as a pair, except you do not need the doorway. Carry the lower window rail across what was the door opening, to provide support for the plywood covering. You can move the upright to take the place of the window upright, or you could increase the widths of the windows on that side. At the rear edge, take the covering plywood over the frame 2 inches to allow an overlap on the back.

Place the back over the two sides and extend it to form the rear of the porch (FIG. 5-10A). Arrange one upright between two windows or you could have two uprights equally-spaced between three windows (FIG. 5-8B).

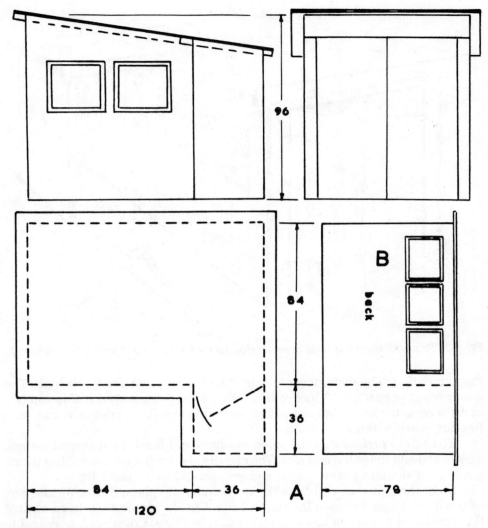

FIG. 5-8. *Suggested sizes for the large-door workshop.*

Make the height of the back to match the adjoining parts of the sides and bevel the top edge to match the slope of the roof (FIG. 5-10B). When you nail on plywood, cut it level with the sides. At the top, leave enough plywood to trim later, since it must reach the roof and since you must notch it around rafters (FIG. 5-10C). When you assemble the building, cover the plywood corners with strips (FIG. 5-10D).

Although the back will fit over the sides, the front is made to fit between them. With the large area cut out for the doors, it needs all the support it can get. Overlapping the sides gives some stiffening. Diagonal bracing above the doorway provides additional stiffening.

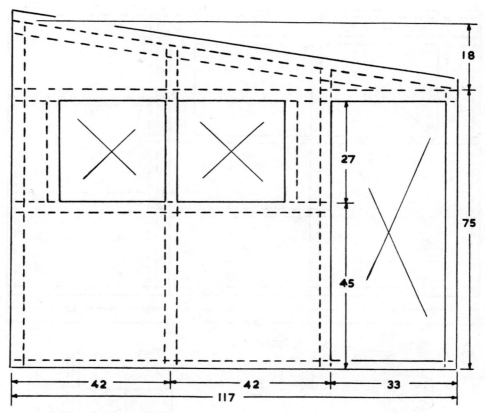

FIG. 5-9. The side of the large-door workshop.

Make the front as shown in FIG. 5-11A. Although it is shown 78 inches across the frame, check this measurement with the back as a guide, allowing for the actual thicknesses of side framing and plywood. The overall width at the front will be the same as that over the sides where they meet the back. Similarly, check the height against the sides, allowing for the bevelled top to match the slope of the roof (FIG. 5-11B). Make sure the frame parts are straight, particularly those which come around the doorway. Arrange strong joints and have the assembly square when you fit the diagonal struts in the top space (FIG. 5-11C).

Cover the framework with plywood. If necessary, pull the side framing of the doorway straight as you fit the plywood. At the sides, let the plywood extend 1 inch to overlap the building side, and allow a filler strip to go in during assembly (FIG. 5-11D). At the top, the plywood will have to reach the roof and you will have to notch it around roof rafters (FIG. 5-11E) as you did on the back.

Make the side of the porch the same size as the main side which includes the door, so it has the same height and top slope. Extend it so it gives good protection to the door. Frame it the same way as you framed a side, with enough

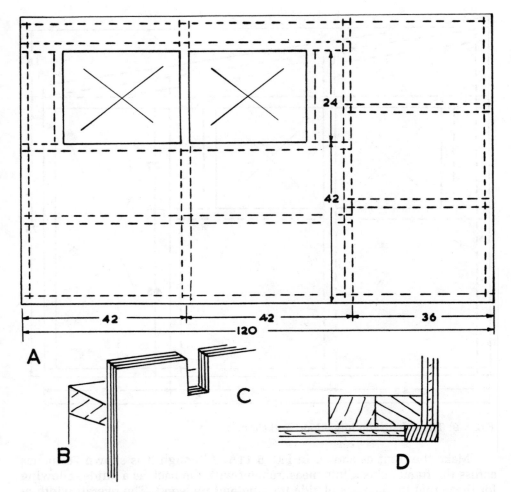

FIG. 5-10. Back of the large-door workshop.

plywood left at the top for trimming after you add the rafters (FIG. 5-12A). It will fit against the back in the same way as the main sides.

Make a front to the porch, framed and covered with plywood, to go across above the door level (FIG. 5-12B). You could put a bar across at ground level, but if you bolt the porch side down, that should be unnecessary. Cover the front edge of the porch side with a strip (FIG. 5-12C). Round its outer edges.

Make the porch door with two pieces of plywood with framing between them (FIG. 5-12D) and thin strips covering the edges (FIG. 5-12E). Fit the framing to one piece of plywood with waterproof glue and fine nails, then add the other plywood. Widen the internal framing, if necessary, where hinges and a lock or latch will come. Put strips around the doorway edges and stop strips on them (FIG. 5-12F).

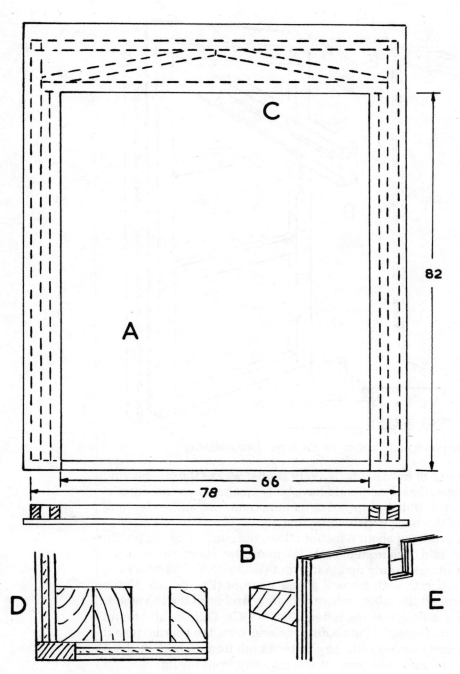

FIG. 5-11. Front details of the large-door workshop.

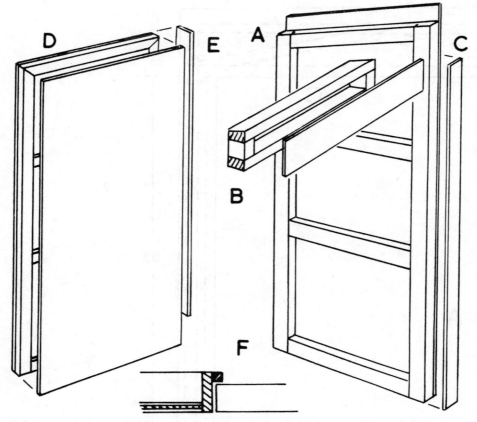

FIG. 5-12. *Details of the porch and its door for the large-door workshop.*

Make and fit the pair of front doors in a very similar way as the other doors. Allow for stouter hinges. One door should have bolts up and down and the other door should lock to it. If draughtproofing is important, put an overlapping strip on the inside of the door which shuts first.

The roof should overhang about 6 inches all around, including the porch. Support it with rafters laid from front to back. Put one rafter directly above each side and space three others evenly across the roof (FIG. 5-13A). Put matching short rafters over the porch, with a board across its front (FIG. 5-13B). Notch the plywood walls around the rafters where necessary and level the top edges.

Lay plywood or boards across the rafters (FIG. 5-13C). Thicken all around the edges with strips underneath. Put roofing material over the boards, turn it under and nail it securely underneath. Lay battens from front to back at about the same spacing as the rafters to prevent the covering from lifting.

At the front, make a fascia board to cover the roof and rafter ends (FIG. 5-13D). You should not need one at the back, but if you do, put one there, for the sake of appearance. Keep it low, so it will not interfere with the run-off of water.

Materials List for Large-door Shop

Sides

2 uprights	2	× 3	×	96
2 uprights	2	× 3	×	80
2 uprights	2	× 3	×	86
2 uprights	2	× 3	×	90
5 rails	2	× 3	×	120
1 rail	2	× 3	×	86
2 window posts	2	× 3	×	32

Back

4 uprights	2	× 3	×	80
2 rails	2	× 3	×	122
3 rails	2	× 3	×	86
2 rails	2	× 3	×	40
2 window posts	2	× 3	×	32
2 corners	1	× 2	×	80

Front

2 posts	2	× 3	×	96
2 posts	2	× 3	×	88
3 rails	2	× 3	×	80
2 diagonals	2	× 3	×	42
2 corners	1	× 2	×	96

Side door

2 sides	1	× 2	×	72
4 rails	1	× 2	×	33
2 panels	72- × -33-× -½ plywood			
2 edges	½	× 3	×	72
2 edges	½	× 3	×	33

Cladding

Shiplap boards 1- × -6 or ¾ plywood (approximately)

Front doors

4 sides	1	× 3	×	84
8 rails	1	× 3	×	36
4 panels	36- × -84- × -½ plywood			
4 edges	½	× 3	×	84
4 edges	½	× 3	×	36

Door and window edges

4 uprights	1	× 4	×	84
1 front top	1	× 4	×	68
1 side-door top	1	× 4	×	33
8 window sides	1	× 4½	×	28
2 window tops	1	× 4½	×	66
2 window sills	1	× 5½	×	66
Stops from seven	1	× 1	×	82
16 window frames	1½	× 1½	×	30

Roof

5 rafters	2	× 3	×	136
2 rafters	2	× 3	×	48
Covers	1- × -6 boards or ¾ plywood			
3 edges	1½	× 1½	×	136
1 edge	1½	× 1½	×	100
2 edges	1½	× 1½	×	48
1 fascia board	1	× 6	×	100
1 fascia board	1	× 6	×	48
4 battens	½	× 2	×	130

Porch

2 uprights	2	× 3	×	82
4 rails	2	× 3	×	48
2 front rails	2	× 3	×	40
1 front edge	1	× 3	×	80

You can use windows which are fixed or which open. Make them similar to the ones made in the last project. Start by lining the window openings with strips at the sides and top (FIG. 5-4A) and use wider sills at the bottom (FIG. 5-4B). For windows which open, arrange stops and make the window frames as described for that project (FIG. 5-4C,D,E). You can make fixed windows in the same way, attaching the windows to the stops. An alternative would be to

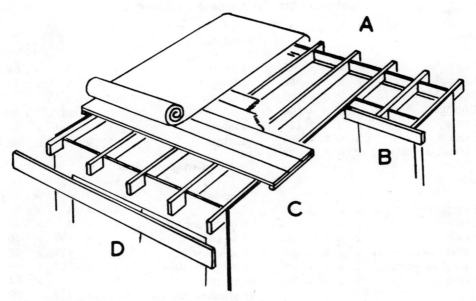

FIG. 5-13. *The roof of the large-door workshop.*

putty the glass directly against the stop strips. It would look better if you would use rabbetted strips instead of the stop strips.

STUDIO

Anyone practicing an art form needs good, all-around light which doesn't glare. This fact applies to three-dimensional carving and sculpture as well as to painting. There should be plenty of windows that let in light where needed, as broadly as possible, so there is no glare and harsh shadows are not cast.

Sloping windows will pass light without glare better than upright windows and they will spread the natural illumination. An artist usually wants one wall without windows. The size of a studio will depend on the work to be done and how many people are to be accommodated. The studio in FIG. 5-14 is designed for a single worker on projects of only moderate size. You can use the same construction for a studio of a different size. The suggested sizes are for an 8 foot square floor space and the same size maximum height (FIG. 5-15). Lined walls and roof are advisable. The smooth interior, painted a light color, will help to disperse lighting evenly. Although you might use a concrete base, a wooden floor over it would be comfortable and kinder to dropped tools. Cladding is assumed to be shiplap boarding, but you could use plywood. The 8-foot-square size makes for economical use of standard plywood sheets.

Start with one side which has a door (FIG. 5-16A). You can use any of the usual framing joints at most places. Where the sloping and vertical fronts join, use a halving joint, with screws both ways (FIG. 5-16B). Cover with boards, cut level at the back and front edges, but with enough left at the top to trim to the same height as the rafters (FIG. 5-16C).

FIG. **5-14.** *This studio gets good light from sloping windows.*

Make the opposite side identical, but instead of a doorway, you might allow space for an opening window, the same width as the door (FIG. 5-16D). Cover with boards in the same way as the first side.

The back is a simple, rectangular frame (FIG. 5-17A). Divide the width into four and put a central rail across. Check the height against the matching parts of the sides, and bevel the top frame member to match the slope of the roof. At the top, let the covering boards project about 3 inches. When you assemble the studio on-site, notch the covering boards for the rafters and trim level with their top surfaces. At the sides, allow the covering to extend by the width of the end uprights, then when you assemble, the board ends will overlap and you can put a square strip in the corner (FIG. 5-16E).

Make the lower front (FIG. 5-17B) to match the back width. The covering boards are level at top and bottom edges, but they extend at the sides in the same way as on the back.

Fully glaze the upper front. Divide it into five glass panels (FIG. 5-18A). The overall length should be the same as the lower front, and the height must match the sloping parts of the sides. All of the parts are rabbetted—⅝ inch deep and ⅝ inch wide should be enough. The outside parts have rabbets on one edge (FIG. 5-18B). The intermediate pieces have rabbets on two edges (FIG. 5-18C).

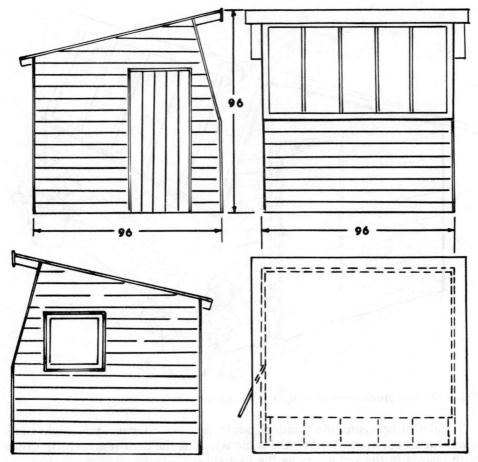

FIG. 5-15. *Suggested sizes for the studio.*

It is possible to dowel parts together, but the best joints are mortises and tenons (FIG. 5-18D). Treat the corners similar to the intermediate joint shown, but reduce the width of the tenon at the outside. Do not fit the glass until after you have erected the building.

Arrange a sill on the lower front for the upper front to fit over (FIG. 5-18E). You can extend the sill inwards to make a shelf (FIG. 5-18F). It would be difficult to waterproof the joint between the upper frame and the sill with glue. It is better to embed it in jointing compound. Cover the ends of the upper front with strips over the edges of the studio ends, when you assemble it.

Make the opening window in the end the same way as described earlier (FIG. 5-4). Make the door with vertical boards and ledges and braces, as described (FIG. 5-5). If you do not include a window, put a second brace across. Braces should slope up from the hinged side. Line the doorway and arrange stop strips in the way described earlier for the door you are using.

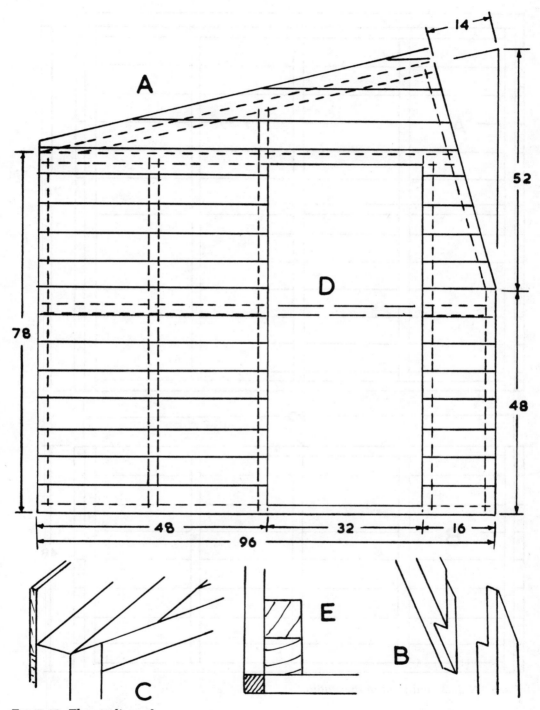

FIG. 5-16. The studio end.

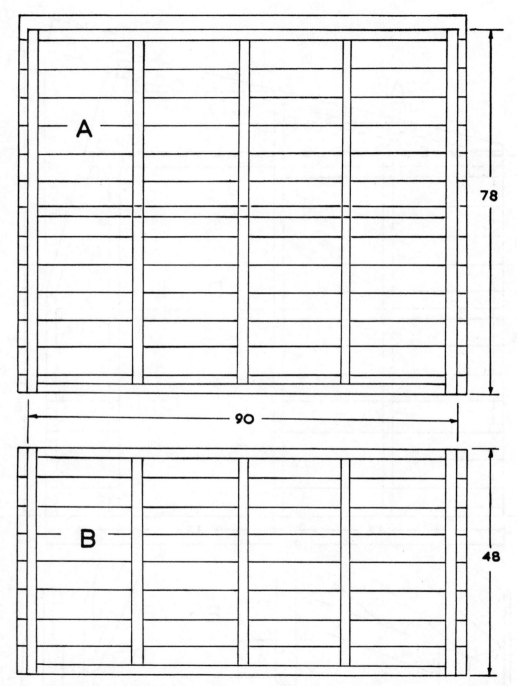

FIG. 5-17. Back and front of the studio.

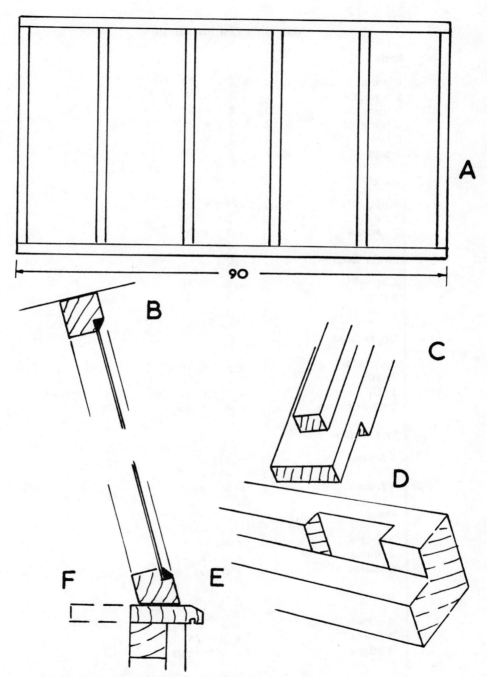

FIG. 5-18. Glazing arrangements for the studio.

Materials List for Studio

Ends

6 uprights	2	× 2	× 80
2 uprights	2	× 2	× 90
2 uprights	2	× 2	× 50
2 uprights	2	× 2	× 56
6 rails	2	× 2	× 98
2 tops	2	× 2	× 98

Back

5 uprights	2	× 2	× 80
3 rails	2	× 2	× 92
2 corners	1	× 1	× 80

Lower front

5 uprights	2	× 2	× 50
2 rails	2	× 2	× 92
2 corners	1	× 1	× 50

Upper front

6 uprights	2	× 2	× 56
2 rails	2	× 2	× 92
1 sill	1	× 4½	× 92
2 cover pieces	1	× 5	× 56

End window

3 frames	1	× 3½	× 34
1 sill	1	× 4½	× 34
4 frames	2	× 2	× 34

Door

3 ledges	1	× 6	× 32
2 braces	1	× 6	× 40
6 boards	1	× 6	× 78

Roof

5 rafters	2	× 3	× 116
1 fascia	1	× 6	× 116
4 edges	1½	× 1½	× 116

Covering: boards   1- × -6 or ¾ plywood

Cladding

Shiplap boards   1- × -6 or ¾ plywood (approximately)

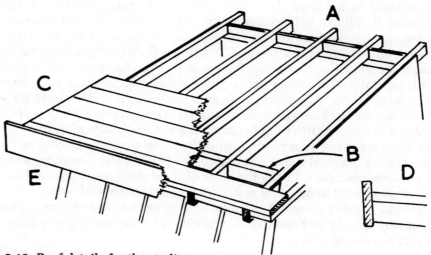

FIG. 5-19. *Roof details for the studio.*

Make the roof with rafters laid from back to front (FIG. 5-19A). Let the rafters project about 6 inches at back and 12 inches at front. Notch the plywood at the back. At the front, fill the gaps between the rafters with 2-inch × 3-inch pieces, level with the front of the glazed part (FIG. 5-19B).

Cover the rafters with boards across (FIG. 5-19C) to give a 6-inch overhang at the sides. Thicken all edges with strips underneath. Bevel the front so it is vertical (FIG. 5-19D). Turn the covering material under and nail underneath to the strips. Put battens on top from front to back.

At the front, put a fascia board across (FIG. 5-19E). Do not make it too deep or it will restrict light in the windows. You could give it a decorative shape if you wish.

When you have completed assembly, you can line the walls and under the rafters with plywood or particleboard. If you mounted the building on a concrete base, there could be a wooden floor. Place the boards over the bottom framing parts and lay stiffeners across underneath at about 18 inch intervals. Fit the floor before lining the walls.

SUBSTANTIAL WORKSHOP

If you need a small building to use as a shop for year-round work, or you want to install several machines or other heavy equipment, the construction ought to be more substantial than many shops which have sectional construction. The sectional-construction building are satisfactory for lighter or occasional use. If climatic extremes affect the contents, insulate the building adequately. This insulation also will improve your personal comfort. It is probable that you will need to use plenty of electricity, which would include the accompanying switch-gear and fuses or cutouts. Electricity is installed more safely in a building such as a house rather than a temporary shed.

Such a substantial shop obviously will be more costly than one of lighter construction. Building it will involve more work, mostly on-site, but if you want a long-lasting shop of the best construction, this shop is it. Line and insulate the walls. You also might line the roof. A wooden floor covers a concrete base. You can place a simple, single door in one end or you can enlarge it to double doors or you might fit double doors at the other end. Even if your normal activities do not need wide doors, make sure your door is wide enough for you to pass the largest machine you wish through it. You can place windows in one side only, with more natural light coming from the opposite roof, or you can arrange windows to suit your needs, during the initial planning stage. You probably will have a bench at the window side, for hand work, assembly, and the use of portable machines. Check the sizes of fixed machines. Locate them so you can move around them safely and there is clearance to work with sheet and long material. Think of storage. You might arrange racks inside, although a lower lean-to shelter could cover racks along one side. This design might be valuable for natural seasoning of wood.

Available space and access to it might control sizes. For example, it is assumed that the building is 10 feet wide, 14 feet long, and that there is a lean-to store 4 feet wide (Fig. 5-20). The height is 8 feet to the eaves. One wall and

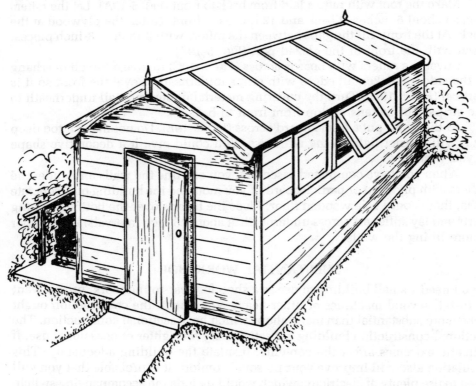

FIG. 5-20. This workshop has an extra-strong construction to withstand the activities of many occupations and crafts.

one end are solid. The solid wall gives firm and adequate attachments for machines which you need to mount on the wall, and plenty of space for shelves and cabinets. Windows and doors take away a surprising amount of wall space, and you will have to weigh the value of cutting through these walls against their use when left solid.

The building will be fairly heavy. The equipment and stock you add will represent more weight. If you add machines with plenty of cast iron in them, the total weight on the base might be more than you first visualized. If an inadequate base allows part of the building to sag, rectification at a later date will be difficult, if not impossible. Fortunately, you can prepare a concrete base of sufficient strength almost as easily as a thinner, poorly-supported base. Dig deep enough to fill under the base with compacted stone, then lay more than 4 inches of concrete on top of it, taking it about 6 inches outside the building area and keeping the top surface above the surrounding ground.

Although it is possible to put the wooden floor directly on the concrete base, it is better to raise it. Several ways are possible. You could cement bricks or concrete blocks in place. You might use railroad ties. If you use new wood, it should be pressure-impregnated with preservative. This type wood would be advisable for the floor framing as well. Use 2-inch × 4-inch wood, or larger. You can arrange the supports all around, but it would be better to have them only under the floor joists, so there is ventilation (FIG. 5-21A). Spike or bolt the wood to the concrete. Alternate bearers might be full-length. You could place short pieces intermediately, depending on the stiffness of the floor. Put polyethylene, or other plastic sheeting, between the wood and concrete to reduce the amount of moisture meeting the wood.

You could put the floor joists in position and nail full-length floor boards to them, but if you prefer to put the floor together away from its final position, it could be in two parts which you would bolt together (FIG. 5-21B). Boards might be plain, but tongue-and-groove boards will prevent gaps, if there is shrinkage. Alternatively, use particleboard or thick plywood. Polyethylene sheeting between the floor boards and the joists (FIG. 5-21C) will act as a vapor barrier. For maximum floor insulation, sandwich insulating foam between the floor boards and plywood from below (FIG. 5-21D).

The floor settles the shape of the building, so take care to get it and its supports square. Compare diagonal measurements. Before proceeding with the walls, it is advisable to fix fine, metal mesh at open ends (FIG. 5-21E) to keep out leaves and vermin, without restricting ventilation.

You can do much of the assembly of sides and ends flat on the floor. The square corners of the base will serve as guides in squaring wall assemblies. Make the walls so the outside of a frame comes level with the outside of the floor and take the siding a little way below the floor level (FIG. 5-21F). As you mark out the frames, allow for the corner joints. Bolt the 2-inch × 3-inch uprights together and cover the siding with an upright corner strip (FIG. 5-21G). Include a fillet (FIG. 5-21H) to support the lining. Joints between frame parts might be the same as in earlier buildings, preferably with open mortise-and-tenon joints

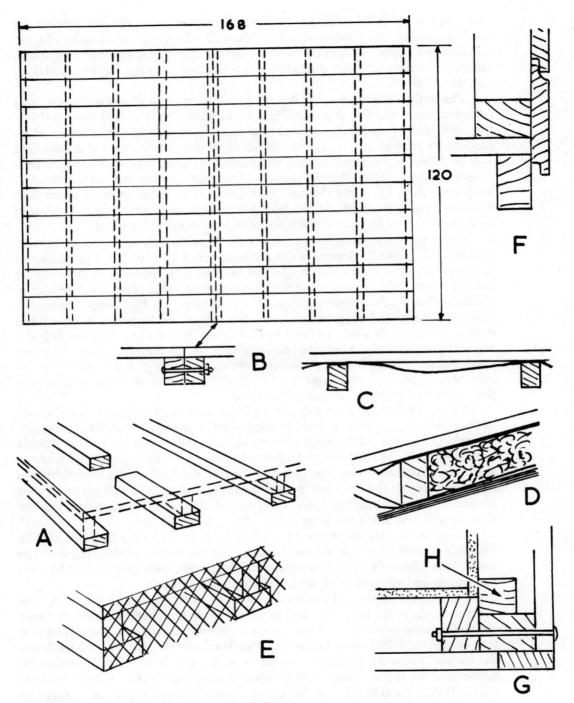

FIG. 5-21. Floor details for substantial workshop.

at the corners. Shallow notches or halving joints are suitable elsewhere. Diagonal-strut-sway bracing is advisable in a building of this size, particularly if it will be exposed. This bracing resists wind loads and relieves the skin material of racking loads under strain that it might otherwise get. Covering with plywood sheets might provide stiffness without the need for diagonal struts, but with shiplap siding, they are advisable.

The two sides are rectangular frames the same length as the floor. They are shown 8 feet high, but if you will be covering with standard plywood sheets, you might reduce the height a little to allow for the sheets to project over the edges of the floor. For shiplap-board covering, you might keep the height to 8 feet, which will then suit the standard sheets you use for lining.

The closed side (FIG. 5-22A) has uprights at 24-inch intervals, and two equally-spaced rails are halved to them. At the end, fit the cleats (FIG. 5-21H and 5-22C), if you are going to line the walls after assembly. The four sway-bracing diagonals should fit closely and you should securely nail them to the framing. You can arrange this bracing easily by putting blocks in the corners and cutting the diagonal ends to fit against them (FIG. 5-22D).

Cut the cladding boards level at the ends. Along the bottom edge, allow a projection to overhang the floor (FIG. 5-21F). You can finish the top board level with the top of the frame, or you might prefer to fit it after you have assembled the building as you will have a gap to fill after you have boarded the roof, and you might continue upwards with a wide board. If you cut the boards flush with the frame, you can put filler pieces on top later.

The side with the windows has the same overall size (FIG. 5-22B). You can arrange the window opening to suit your needs, but the arrangement shown has three windows about 36 inches square, 48 inches above the floor. If you plan a different window layout, have sufficient uprights not more than 36 inches apart to support the cladding. Fit cleats and arrange sway bracing in the same way as on the other side. Put the cladding boards on in the same way. At the window openings, cut the boards level. You can cover their edges when you frame the windows.

The two end frames are almost the same (FIG. 5-23A). Make the closed end like the door end, but take the central rail right across (FIG. 5-23B). Although an overall width is given, it is important that the ends fit between the sides (FIG. 5-21G,H). Check on the floor that the ends will hold the sides the correct distance apart for you to take the cladding down outside the floor. Also check that the eave's height matches the sides.

Assemble the frames squarely. Halve the sloping tops to each other and to the uprights. The doorway is 42 inches wide and the top is 84 inches above the floor. You could modify these sizes at this stage, if you wish. Include sway bracing similar to that on the sides.

Three 2-inch × 3-inch purlins are on the edge at each side. In this building, there is no separate ridge. Instead, the top purlins are fairly close to the apex, to support the roof boards there. Arrange lower purlins close to the eaves and the others midway. Put supporting cleats on the frame (FIG. 5-23C).

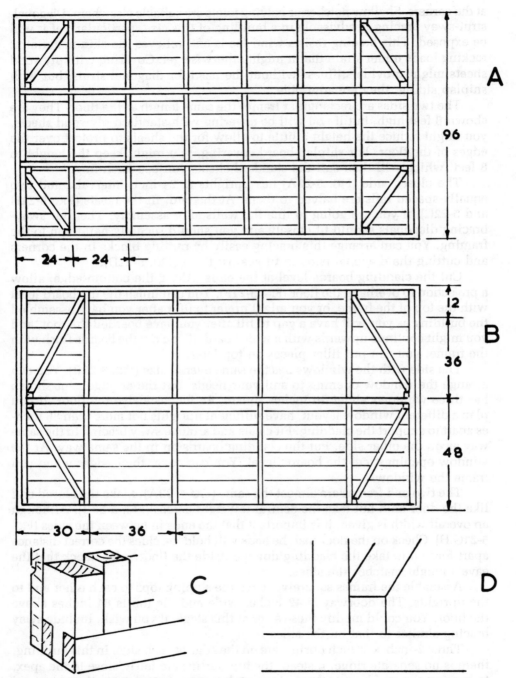

FIG. 5-22. Walls of the substantial workshop.

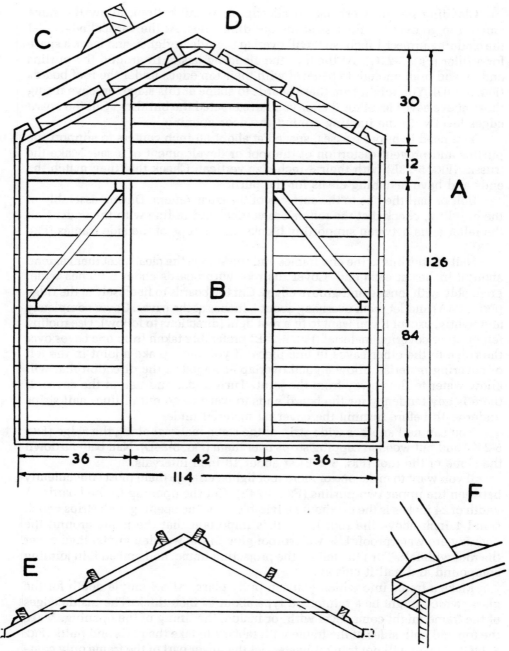

FIG. 5-23. End and roof truss of the substantial workshop.

Cladding has to overhang on all edges. At the bottom, allow the same amount to go over the floor as at the building sides. At the vertical edges, let the cladding project 1 inch, so it will overlap the side uprights, and leave a space for a filler (FIG. 5-21G). At the top, the cladding has to fit around the purlins and extend high enough to be level with their top edges under the roof boards (FIG. 5-23D). You might trim these boards to shape at this stage, or leave fitting those above eaves level until after you have erected the building. Cut the board edges level with the framing around the door opening.

You need two roof trusses, spaced at about 56-inch centers to support the purlins and prevent distortion of the roof or development of a sag. Make the trusses (FIG. 5-23E) with their 3-inch size vertical. Check that they match the ends and have matching cleats for the purlins.

Bolt or nail the ties to the surfaces of the truss rafters. During assembly of the building, check that the rafters are vertical and in line with the ends. Nail the rafter ends between supporting blocks on the tops of the side frames (FIG. 5-23F).

Nail the purlins to the end frames and trusses via the cleats. Let them extend about 6 inches at each end. Cover the roof with boards about 6 inches wide, preferably with tongue-and-groove edges. Cut the boards to fit closely at the ridge (FIG. 5-24A) and let them overhang about 5 inches at the eaves. Cover completely and tightly, except if you want to fit a roof light (directions to follow). Use roofing felt or other covering material (FIG. 5-24B) preferably taken from one eaves over the ridge to the other eaves in one piece. If you must make a joint in the felt or covering material, allow a good overlap arranged in the direction that will allow water to flow away from the joint. Turn under and nail at the eaves. If there is any tendency for the board ends to warp or go out of line, nail strips underneath before turning the covering material under.

You can nail another wide strip of the same material along the ridge (FIG. 5-24C) and put wooden capping strips over them (FIG. 5-24D). Nail battens down the slope of the roof (FIG. 5-24E) at about 18 inch intervals.

If you want to make one or more roof lights, arrange them most conveniently between the upper two purlins (FIG. 5-24F). Cut the opening in the boards. A width of 24 inches is the maximum advisable. Line the opening with strips which stand 1 inch above the roof level. It is important that the joints around the opening are waterproof. Use waterproof glue in the wooden joints, then cover the roof with felt. Turn the felt up the projecting frame, and embed it in jointing compound and nail it closely.

Make a frame into which you can putty glass. Allow ample depth for the glass, which might be a reinforced type about ¼ inch thick. The inside edges of the frame might come level with, or inside, the lining of the opening. Make the top and both sides of the frame with rabbets to take the glass and putty (FIG. 5-24G). So you will not trap rainwater, let the lower part of the frame only come under the glass, which might project there slightly (FIG. 5-24H).

Mount the frame in position by gluing and screwing it to the lining pieces. Complete painting this woodwork before embedding the glass in putty. If you have doubts about the putty being able to prevent the glass from slipping, nail

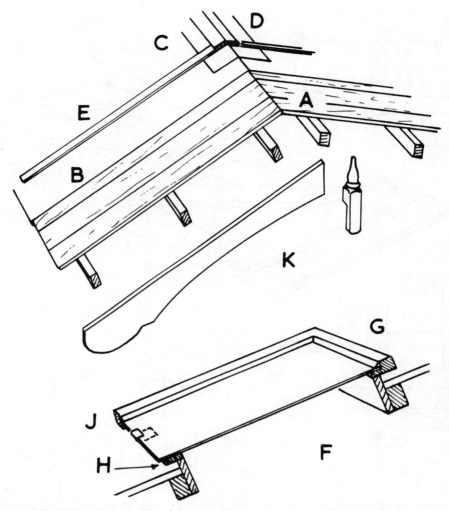

FIG. 5-24. *Roof, bargeboard, and roof light details for the substantial workshop.*

two sheet-metal pieces to the lower part of the frame and bend their ends around the glass (FIG. 5-24J). Fit bargeboards to both ends of the roof. They could be straight or you can decorate them in the ways described for some earlier buildings. A different decorated pattern is suggested in FIG. 5-24K. With this pattern goes a turned final on a square part notched over the apex of the meeting bargeboards.

Line the door opening (FIG. 5-25A) with strips at sides and top. Round the projecting edges, and at the floor continue the lining pieces over the floor boards.

Ledge and brace the door in the usual way, except there is a covering piece at the top, with the brace immediately below it. At the bottom, the door overlaps the floor, as weatherproofing. Put stop strips in the sides of the doorway (FIG. 5-25B), but leave space at the top to clear the covering piece.

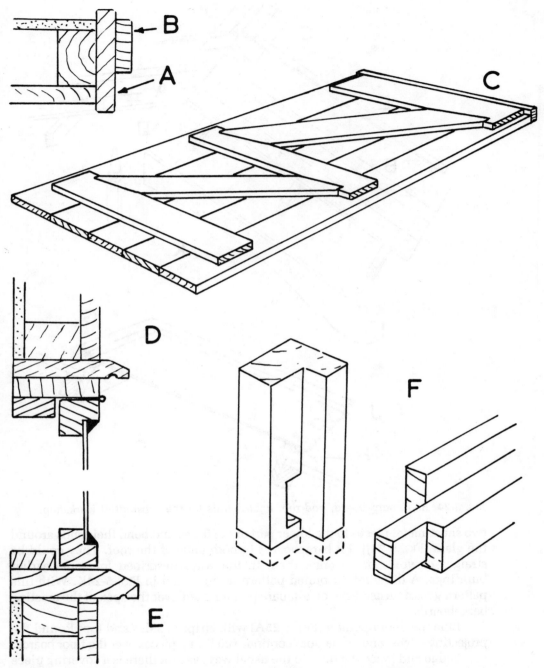

FIG. 5-25. *Door and sections of the window parts of the substantial workshop.*

Material List for Substantial Workshop

Floor

10 joists	2 × 3 × 125		
20 boards	1 × 6 × 86		
	or equivalent area		

Sides

12 uprights	2	×	3	×	98	
7 rails	2	×	3	×	170	
1 rail	2	×	3	×	120	
4 sway bracings	2	×	3	×	36	
4 sway bracings	2	×	3	×	56	
6 fillets	2	×	2	×	34	
4 fillets	2	×	2	×	48	

Ends

4 uprights	2	×	3	×	98	
4 uprights	2	×	3	×	118	
5 rails	2	×	3	×	116	
2 rails	2	×	3	×	38	
4 sway bracings	2	×	3	×	48	
4 top rails	2	×	3	×	66	

Trusses

4 rafters	2	×	3	×	66	
2 ties	2	×	3	×	100	

Cladding

Cladding	1- × -6 shiplap boards or equivalent plywood
Lining	½ plywood or particleboards

Roof

6 purlins	2	×	3	×	180	
60 boards	1 × 6 × 72 or equivalent area					
4 bargeboards	1	×	8	×	76	
2 finials	3	×	3	×	20	
4 roof-light linings	1	×	5	×	25	
3 roof-light frames	1½	×	3	×	30	
1 roof-light frame	1	×	3	×	30	
18 battens	½	×	1½	×	70	
2 cappings	1	×	3	×	180	

Door

2 linings	1	×	6	×	86	
1 lining	1	×	6	×	44	
5 boards	1 × 9 × 86 or equivalent area					
3 ledges	1	×	6	×	40	
2 braces	1	×	6	×	42	
1 top	1	×	2	×	42	
2 stops	1	×	3	×	84	

Windows

6 linings	1	×	6	×	38	
3 top linings	1	×	7	×	38	
3 top-hinge rails	1	×	5	×	38	
3 sills	1½	×	7½	×	38	
12 window moldings	2	×	2	×	38	

Use tongue-and-groove boards for the door. At the top, glue and screw the covering piece on (FIG. 5-25C). Make the ledges short enough to clear the stop strips. Fit the braces sloping up from the hinge side. Arrange hinges and fasteners over the ledges.

If you are going to line the building, you might do it before you fit the windows, so their framing can cover the lining boards, which might be plywood or particleboard. Include polyethylene sheet as a vapor barrier, if you wish. You can include insulating material in the space.

At the roof, close any spaces around the edges with cladding carried up to the roof boards or with pieces on top of the side frames. Nail lining material to the undersides of the purlins.

At the sides of the window openings, fit lining similar to that at the sides of the doorway. You can line the top in a similar way, but it would be more weatherproof if a piece extends and there is another strip below it for the window hinges (if you wish to make them open) (FIG. 5-25D). At the bottom, make a wider sill to shed water (FIG. 5-25E).

Make the windows with rabbetted strips, with mortise-and-tenon joints at the corners (FIG. 5-25F). Fit hinges at the top and a stay and fastener at the bottom, if the window is to open. Screw other windows into their openings, preferably embedding them in jointing compound.

As the shop floor is above the surrounding concrete base, you might make a wooden or concrete step at the doorway. If there is to be a storage lean-to along one side, you can make it with a roof similar in construction to the main roof, with open-frame supports and racks underneath.

GREENHOUSE

A greenhouse is a gardener's workshop. Besides growing some things completely and starting others before putting them in the ground, it is a shelter to work in when conditions outside are uncomfortable and it is a place to plan new horticultural projects. It extends the gardener's season for his occupation or hobby. For an enthusiastic gardener, a greenhouse is almost essential. A second stage in preparing plants to be put outside is to place them in a cold frame to harden them off. Quite often, that structure is a crude improvisation. In this building, a cold frame is included as part of the same unit, so you can efficiently tackle that stage of gardening (FIG. 5-26).

Construction of a greenhouse can be very similar to that of any other building of the same shape, except you cover large areas of the walls and roof with glass instead of wood. You can prefabricate this building to a large extent, with ends and walls ready to bolt together. You make the roof on-site and glaze after erection, although a careful worker might prefer to fit some glass while the walls are flat on the floor. One problem is flexing, which could loosen putty or even break glass. In most cases, it is better to leave glass until you are satisfied that you have finally, and rigidly, assembled the greenhouse.

The sizes suggested are for a building of modest size (FIG. 5-27A,B), but you can use the same method for other sizes. Most of the structure is made from 2-inch-square wood, with some wider pieces where you need extra strength. Use shiplap boarding outside to near bench level. Above that, use glass. The door is at one end. A shuttered hole provides control of ventilation. Locate this hole low in the door and another one high in the opposite end. Make the cold frame by extending the sides, so you can finish it with access via a lifting top. It does not have any connection through to the inside of the greenhouse. As a result, it is not affected by any heating in the greenhouse and if you put soil into it, it is kept away from the floor of the main building.

The greenhouse could have a wooden floor, but because of water often running about, it would be better to lay a concrete base. The concrete base could go under the cold frame or that might be preferable with a soil base. You can

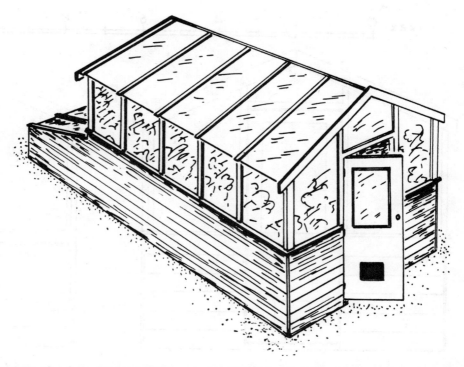

FIG. 5-26. You board this greenhouse to bench height and then build a cold frame onto the end.

cut drainage channels in the concrete and lead them under the greenhouse walls to take away surplus water.

The two ends fit between the sides. For an overall width of about 96 inches, you can arrange the uprights to be 30 inches between centers. This spacing gives a suitable width for the door and you simplify glass cutting by making all panes the same width (FIG. 5-28A). Joints in the upper parts of the frames will be without the additional strength of boarding. It is advisable to use open mortise-and-tenon (or bridle) joints, with waterproof glue and drive either nails or dowels across each. The lower joints can be the same or the simpler, halving joints.

Start with the door end (FIG. 5-28B). The bottom rail goes right across. Other short rails mark the height of the doorway and the cladding. At the apex, leave a gap to take the 2-inch × 6-inch ridge standing 2 inches above the framing (FIG. 5-28C), and put a short rail below the opening to brace the rafters and support the ridge (FIG. 5-28D).

Let the cladding project on each side to go halfway over the side posts (FIG. 5-28E). You will fill the space with a strip when you assemble the building. There will be a sill above the cladding (FIG. 5-28F). Since the sill looks best if you miter it around the corner, you might want to install it after you have assembled the walls.

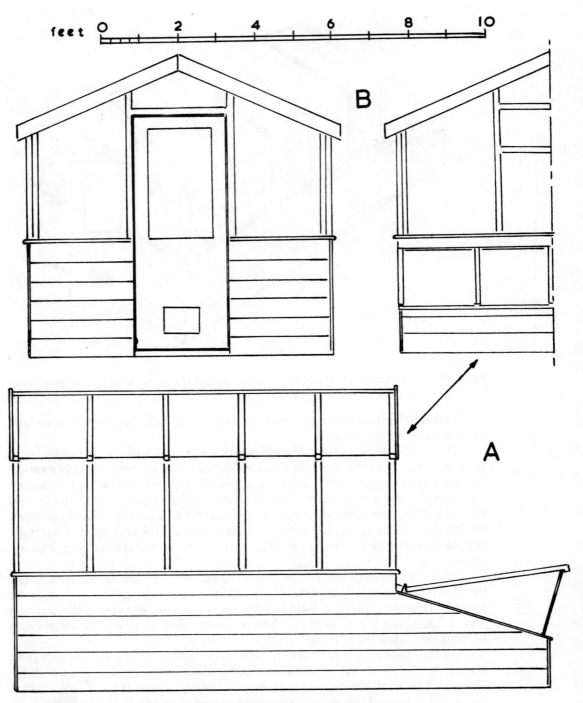

FIG. 5-27. Four views of the greenhouse, showing sizes and proportion.

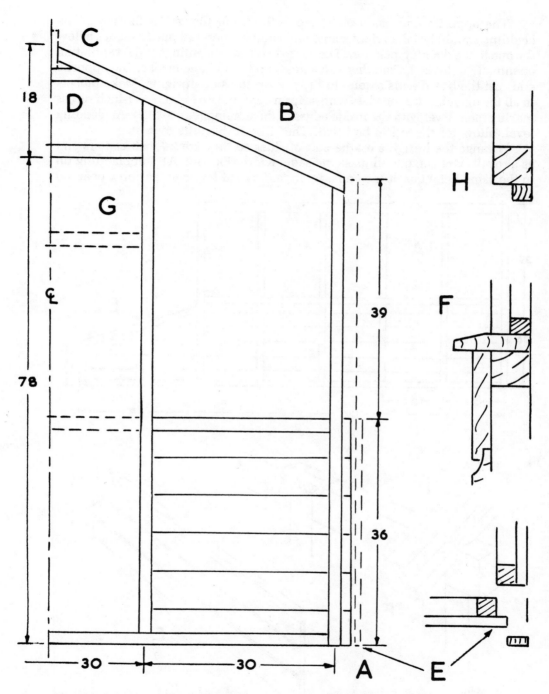

FIG. 5-28. *The door end of the greenhouse and sections through windows and corners.*

The opposite end has a matching outline, but the rail at the top of the cladding should be taken right across and another short rail put 12 inches below the one in the door top position. The space between the railings is the ventilation opening (FIG. 5-28G). Clad that end across, up to the same level as the opposite end. Let the board ends extend to fully cover the side uprights. Make rabbets in all the openings for the glass with strips nailed in. Use ¾-inch × 1-inch pieces making them level with the inside edges of the framing (FIG. 5-28H). At cladding level, allow for the sill to be fitted. The sill will have its own strip.

Arrange the uprights on the pair of sides so they are 24-inch centers and as a result, you can cut all glass the same width (FIG. 5-29A). Put cladding up to the same height as at the ends. Include the cold-frame extensions, boarded

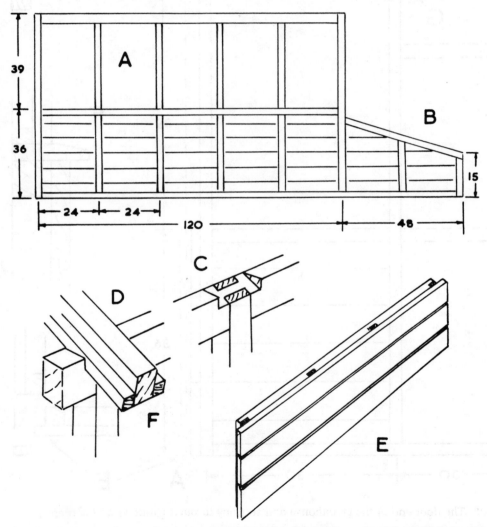

FIG. 5-29. A greenhouse side (A,B), rafter details (C,D) and the end of the cold frame (E).

to the top (FIG. 5-29B). The top member is a 2-inch × 4-inch section. Rafters have to rest on it. Instead of cutting its top to suit the roof slope, leave it square, except for notches at the roof angle where the rafters will come directly over the uprights (FIG. 5-29C,D). After you assemble the building, you can put filler pieces on the squared tops to close the gaps between the wall and the roof glass.

Cut off the ends of the cladding boards level with the end uprights. At the end of the cold frame, make a piece the same width as the greenhouse ends so it fits between the sides (FIG. 5-29E), with cladding to the top. Cut the cladding at the same angle as the sides. Let the ends extend to cover the side uprights in the same way as the door end (FIG. 5-28E).

Assemble all the parts made so far, using ⅜-inch bolts at about 15-inch intervals through the posts. Put filler pieces in the corners of the cladding. Check squareness and fasten the bottom frame members down to the base.

Fit a sill all around. Level the sill inside and taper it outside from just outside the edges of the uprights (FIG. 5-28F). You can extend it a few inches inwards if you want to make it into a shelf. You can make the sill in sections, but you can obtain the neatest finish by making it continuous on the sides and ends. For the strongest construction, notch the sill into the uprights about ¼ inch (FIG. 5-30A). At the corners, extend the sill parts to miter together (FIG. 5-30B). Put rabbet strips in place.

Make the 2-inch × 6-inch ridge to fit in the slots in the ends and be level outside. Mark on it the positions of the rafters, to match the positions of the slots on the tops of the building sides. Prepare the rafters with rabbet strips on each side of the intermediate ones (FIG. 5-29F) and on the inside edges only of the end ones. Using the angle of a building as a guide, cut the tops of the rafters to fit against the ridge (FIG. 5-30C). Make the length of a rafter enough to overhang the walls by 6 inches. Prepare all twelve rafters, so they match.

Nail the end rafters on top of the frame ends (FIG. 5-30D). At the other positions, put a supporting strip underneath (FIG. 5-30E) as you position each rafter. Nail the lower end into its recess. Make rabbet strips with sloping tops to go between the rafters (FIG. 5-30F).

This procedure completes the assembly of all parts that you must glaze. Putty and the alternative compounds do not bond well with untreated wood, so either paint all the woodwork or just the rabbets, so they will dry while you work on other parts.

Glaze the door to the same level as the walls. A ventilation hole is in its bottom panel. The suggested construction has ¼-inch-plywood panels on each side with a frame of 1-inch × 2-inch strips inside (FIG. 5-30G).

Make the door an easy fit in the doorway, with its bottom above the bottom framing strip. The plywood could go to the edges without further protection, but it will be better to cover with ½-inch strips (FIG. 5-30H). Allow for this covering when marking out the plywood panels. Make one plywood door panel first and check its sizes. Glue and fix with thin nails the framing strips around the edges and openings (FIG. 5-30J). When you are satisfied with this panel, make the other panel slightly oversize, so when you have fixed it, you will have true edges.

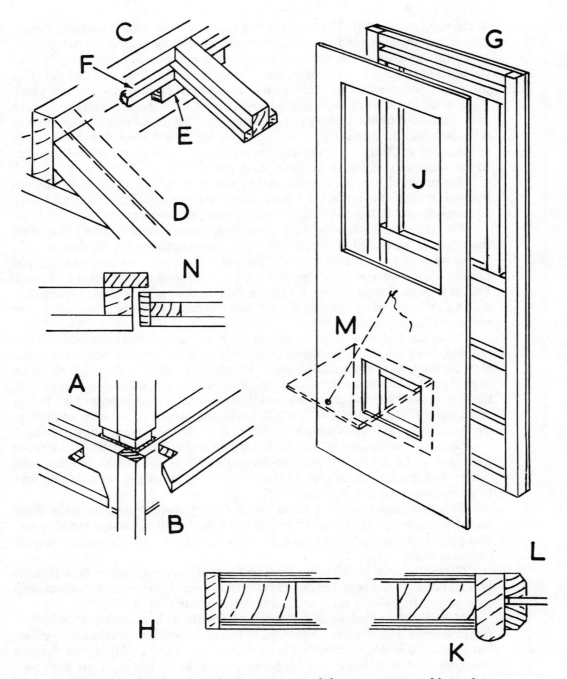

FIG. 5-30. Constructional details of the greenhouse and the way you assemble its door.

Materials List for Greenhouse

Ends

4 uprights	2	× 2	×	78
4 uprights	2	× 2	×	90
3 rails	2	× 2	×	96
7 rails	2	× 2	×	34
4 rafters	2	× 2	×	60
2 sills	1	× 4	×	34
1 sills	1	× 4	×	100

Sides

12 uprights	2	× 2	×	78
2 rails	2	× 2	×	124
2 rails	2	× 2	×	172
2 rails	2	× 4	×	124
4 uprights	2	× 2	×	20
2 strips	2	× 2	×	24
2 strips	2	× 2	×	56
2 sills	1	× 4	×	128

Cold frame

3 rails	2	× 2	×	96
2 rails	2	× 3	×	96
2 covers	1	× 4	×	56
1 cover	1	× 4	×	96
5 glazing bars	2	× 2	×	56

Roof

1 ridge	2	× 6	×	124
12 glazing bars	2	× 2	×	66
4 bargeboards	1	× 6	×	70

Door

2 sides	1	× 2	×	80
6 rails	1	× 2	×	30
2 window posts	1	× 2	×	40
2 edges	½	× 2	×	80
2 edges	½	× 2	×	32
2 window edges	1	× 2	×	40
2 window edges	1	× 2	×	26
2 panels	30 × 78 × ¼ plywood			

Cladding

1- × -6 shiplap boards

Rabbets and fillets   ¾ × 1

Line the window and ventilation openings with strips that are level inside, but project with rounded edges at the front (FIG. 5-30K). As movement of the door, particularly if it is slammed, might loosen putty, it will be better if the door glazing is held between two fillets (FIG. 5-30L). Use jointing compound to embed the glass tightly.

Cover the ventilation opening with fine-mesh wire gauze, to keep out vermin. To control ventilation, hinge a flap on the inside of the door, with a cord to a hook above to regulate the amount of air allowed to pass (FIG. 5-30M). At the other end of the building, arrange a similar flap inside the high opening there, with a cord up to a ring or pulley and down to a cleat within reach.

Strips inside the doorway framing (FIG. 5-30N) will act as stops. Hang the door to swing outwards on three, 3-inch hinges and fit a handle and catch or lock.

Although pieces of glass could be as large as the openings each panel has to fit, you might prefer to use shorter pieces, with the upper pieces overlapping the lower ones, to shed water. This design might be more economical and the first fitting might be handled easily. It also might be easier and cheaper if you ever have to replace a broken pane. Embed the glass in putty, with fine nails (sprigs) holding the glass, then neatly putty over them (FIG. 5-31A).

At the cold frame, cover the top edges with strips (FIG. 5-31B). At the end of the greenhouse, fit a piece of 2-inch × 3-inch wood over the cover strips and bevel it to fit close to the cladding on the greenhouse (FIG. 5-31C,D). Make the lifting cover to fit against this strip of wood overhang with a 6-inch overhang at the bottom. This cover is built up with a 3-inch-wide piece at the top and all other parts are 2-inches square, with ¾-inch × 1-inch rabbet strips (FIG. 5-31E).

Mortise-and-tenon joints are best for the joints to the top rail, but the 2-inch-square bottom piece has to be kept down below the glass level, so water will run off. Halve the ends of the sloping pieces and glue and screw the crosspiece to them (FIG. 5-31F). When you glaze the frame, carry the glass to the edge of the bottom piece (FIG. 5-31G).

The frame with glass will be fairly heavy. Use four strong 4-inch hinges at the top. Arrange a strut to hold the top open.

Finish the ends of the greenhouse with bargeboards. They need not be elaborate, but they will improve appearances by covering the ends of the roof structure (FIG. 5-27B).

CANOE SHED

Canoes and kayaks might not suffer much if left outside, but open storage lacks security, particularly for paddles and other loose equipment. A building will protect these crafts from the weather, as well as allow camping gear and paddles and other associated equipment to be kept secure. The building might have a bench, and you might make repairs and alterations on the spot and in comfort.

Adjust sizes and details of the building to suit canoes, kayaks, sailboards, and similar craft which you wish to store. The canoe shed shown in FIG. 5-32 is intended to provide storage on racks for four canoes or kayaks about 16 feet long. Racks are about 36 inches wide and there is clearance vertically of about

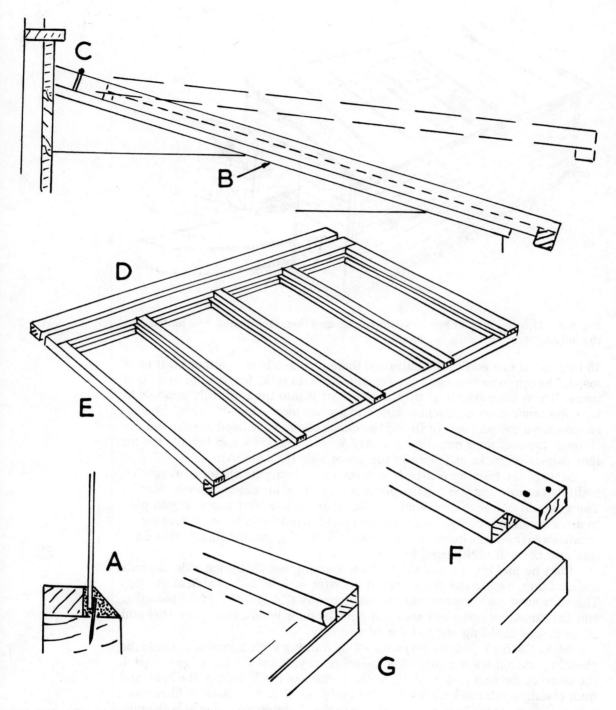

FIG. 5-31. The cold-frame cover and its construction and mounting.

FIG. 5-32. *This canoe shed has racks for canoes and their equipment. The roof admits the only light, so the walls are secure.*

18 inches. At one end of the building, there can be a bench, either built in or loose. The opposite side can have canoe racks with racks for paddles and other items. If you have sailing gear, you can hoist it into the roof. It is possible to have the canoe racks on the high side and get one more canoe in. Then, however, you have the problem of lifting the canoe above your head height and the slope of the roof does not allow for easy access. The sizes and layouts are for four canoes on racks and floor at the lower side (FIG. 5-33A).

Most of the framing is 2-inch × 3-inch wood and the cladding suggested is shiplap boards, but you can use plywood or vertical tongue-and-groove boards. The sides do not have any windows, but there is one in the roof. If you plan to use a bench often, you might wish to put a window in the end. Having no windows in the walls helps security, as breaking in is more difficult. Board the roof and cover it with tarred felt.

Start by making the ends. All of the framing has its 2-inch side towards the cladding, except the rafter, which is better with its 2-inch side at the top. The ends are a pair, except one has the doorway (FIG. 5-34A). Make the other end the same, but put a rail across at half the door height and board that end all over. Cut cladding level at the edges.

Make the back (FIG. 5-35A) with all its framing with 2 inches towards the cladding, except the top, which is the other way. Bevel the top edge to match the slope of the roof (FIG. 5-35B). At the corners of the building, the back and front cladding extends halfway over the end posts (FIG. 5-35C) and a filler piece is fitted in. Cover the back completely, with the cladding extended at both ends.

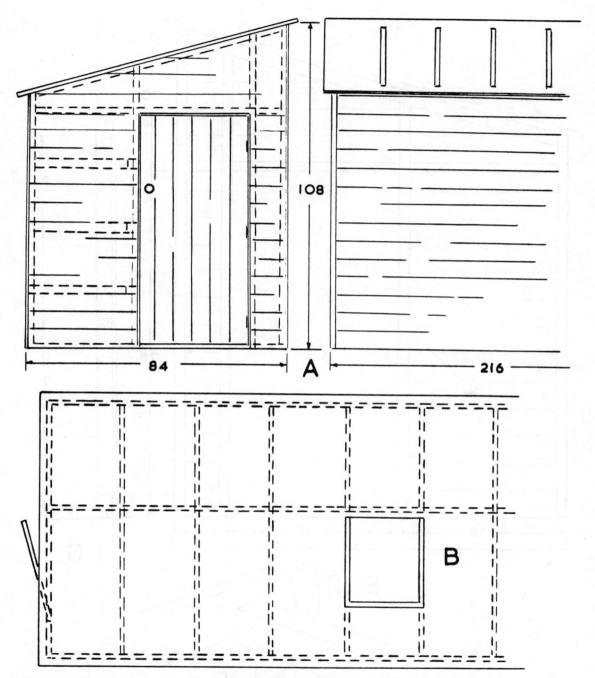

FIG. 5-33. Suggested sizes for the canoe shed, to hold four canoes up to 16 feet long.

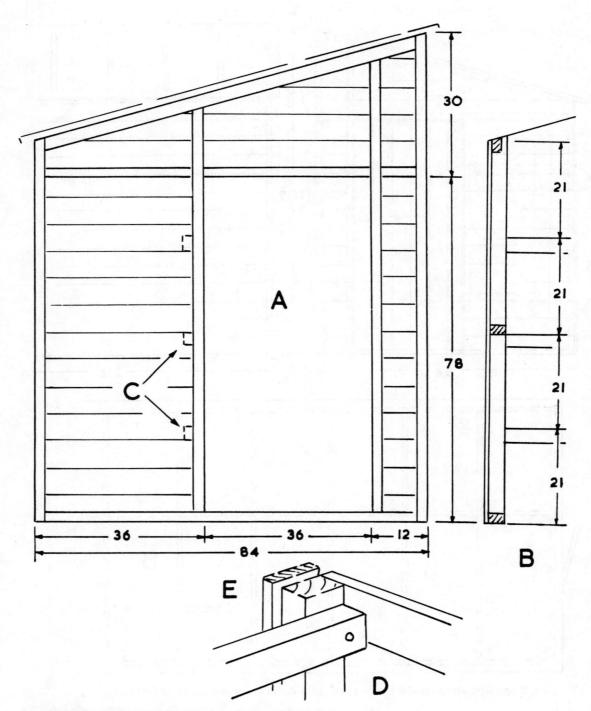

FIG. 5-34. The door end of the canoe shed, and spacing of racks.

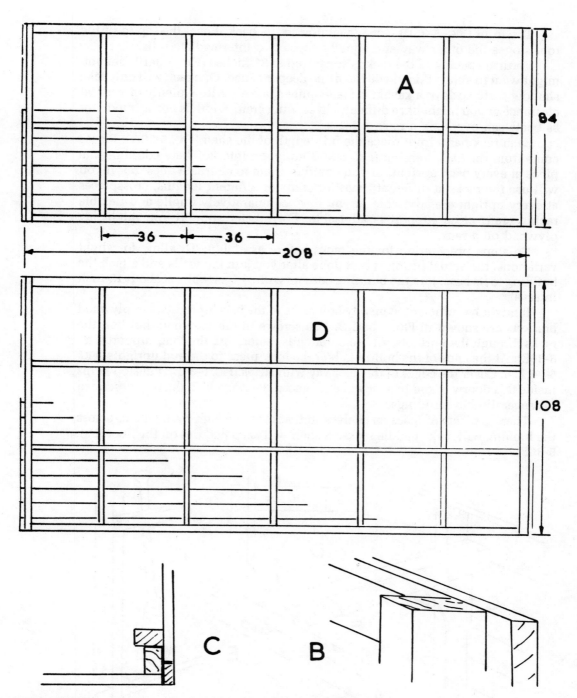

FIG. 5-35. Back (A) and front (D) with corner arrangements (B,C).

The front (FIG. 5-35D) is made similar to the back. Bevel the top to fit the roof slopes the other way and equally space two intermediate rails.

Vertical spacing of the rack is suggested at 21 inches (FIG. 5-34B), but you might want to adjust this, if you have one deeper canoe. One rack will only take shallow surfboards or a kayak, if made quite shallow. Allow plenty of vertical clearance or you might have difficulty in stowing craft. Width is not as important as boats can overhang a rack.

Each rack has a front piece the full length of the shed (FIG. 5-36A). Pieces come from the back framing (FIG. 5-36B) to tenon into it. There could be one piece at every back upright, or you could fit them to alternate ones, but if you will use the racks at different times for crafts of different lengths, crosspieces at every upright are advisable. If you have comparatively fragile and flexible racing craft, close support is advisable. For greater support you might lay plywood on a rack.

Do some preparation for the racks before assembly. To allow for slight variations, the actual fitting is best done after you join the walls and attach the building to its base. Join the building corners with ⅜-inch bolts at about 18-inch intervals.

It might be sufficient to merely bolt racks to the back uprights, but plywood brackets are shown in FIG. 5-36C,D. Use screws in the plywood, but bolt the racks through the uprights. At the outer ends, tenon into the long support (FIG. 5-36E). At the ends of the building, bolt the long piece to the end uprights (FIG. 5-34C,D). Cover the edges of the doorway with strips (FIG. 5-34E). Put stop strips inside this doorway and fit a ledged and braced door, made in the way described for several other buildings.

There is plenty of space for shelves and racks on the high wall. Battens across the framing will make paddle racks. A shelf will keep paddles off the floor (FIG. 5-36F).

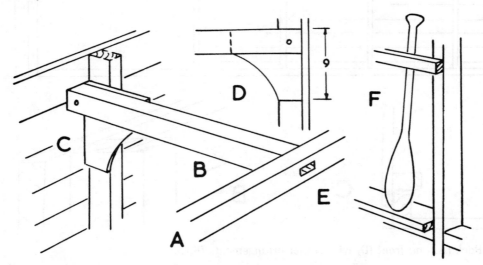

FIG. 5-36. Details of canoe rack construction and a method of paddle storage.

Materials List for Canoe Shed

Ends

2 uprights	2 × 3 × 86
2 uprights	2 × 3 × 110
2 uprights	2 × 3 × 96
2 uprights	2 × 3 × 105
5 rails	2 × 3 × 86
2 rafters	2 × 3 × 100

Back

7 uprights	2 × 3 × 86
3 rails	2 × 3 × 212

Front

7 uprights	2 × 3 × 110
4 rails	2 × 3 × 212
4 fillers	1 × 2 × 110

Roof

8 rafters	2 × 3 × 100
2 fillers	2 × 3 × 26

Roof light

2 frames	1 × 6 × 26
2 frames	1 × 6 × 32
2 sides	2 × 2 × 36
1 top	2 × 2 × 30
1 bottom	1½ × 2 × 30

Door

6 boards	1 × 6 × 78
3 ledges	1 × 6 × 30
2 braces	1 × 6 × 40
2 sides	1 × 5 × 80
1 top	1 × 5 × 32

Racks

3 rails	2 × 3 × 212
15 rails	2 × 3 × 40

Cladding   1- × -6 shiplap boards

Roof   1- × -6 plain or tongue-and-groove boards

Make the roof with rafters on the same level as the tops of the ends, spaced about 24 inches apart (FIG. 5-37A). Notch the rafters into the back and front (FIG. 5-37B), securing them with nails or screws. At the roof light position, halfway along, unless you need it elsewhere (FIG. 5-33B), put pieces across (FIG. 5-37C).

Lay boards lengthwise to cover the roof (FIG. 5-37D). Allow a 3-inch overhang all around. If you need joints lengthwise, make them over rafters. Trim the boards level around the roof light opening. Put a framing of boards around the opening (FIG. 5-37E) to hold the glazed part a few inches above the roof level.

Cover the roof with tarred felt or other flexible roofing material (FIG. 5-37F). Take this over all edges and nail underneath. At the roof light, trim the covering

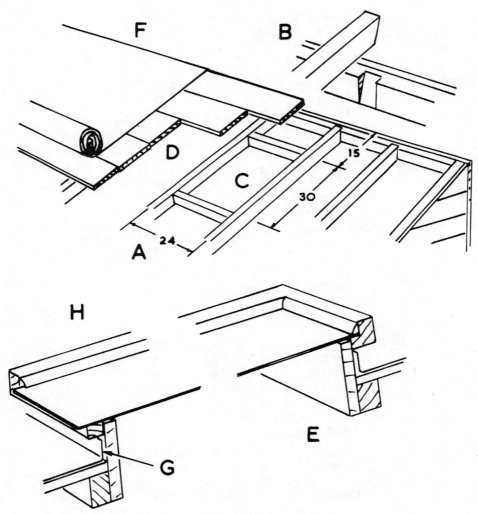

FIG. 5-37. Roof and roof-light details for the canoe shed.

136

so you can turn it up the framing boards (FIG. 5-37G). Nail it there. If there have to be any joints in this vicinity, embed them in a jointing compound.

For the roof light, make a rabbeted frame for the two sides and the top, but at the lower end make a piece to go under the glass only, so water can run off. Have the size of the frame so it will fit over the framing boards and so you can screw to them from inside (FIG. 5-37H). Although you could glaze the window when it is on the ground, it is easier to get the frame closely fitted and the glass accurately puttied in, if you delay the glazing until the roof light is in position.

No bargeboards are shown, but if you want to give the roof a more solid appearance, couple bargeboards to a fascia board along the front edge. If you put a board at the back, make sure it does not interfere with water running off. Lay battens down the roof at about 18-inch intervals, to hold down the covering.

6
DECORATIVE
STRUCTURES

SOME SMALL BUILDINGS MIGHT HAVE PRACTICAL USES, BUT THEY MAINLY ARE intended to be decorative features in a yard, garden, or park. In some cases, it is the foliage growing on or around them that forms the main decoration. The building might be as much a support for climbing trees, creepers, vines, and other foliage, as it is a practical building for storage or just sitting in. Sometimes, what you build is enveloped by natural growth, for which its only function is to provide support. Of course, many functional buildings are decorative in themselves and, even when they are plain, the foliage covering them and around them can provide charm to the general scene. Plants and flowers in pots and boxes can add decoration to what is otherwise a rather austere building.

A decorative building might be a complete structure with walls, door, and roof, particularly if you intend to use it as a storage place. You can build other decorative buildings without a door and they even could be without some or all of the walls and roof. In that case, the woodwork might be there mainly to support trees or bushes chosen for their ability to form a close covering of tightly-packed branches, twigs, leaves, and flowers. What you build is a skeleton on which the growing things fill in to just be decorative or to make a natural wall and roof.

Many names are used for these structures—pergola, gazebo, arbor, summer house, or sun shelter. It is important to remember that if the structure is mainly there to support foliage, it might become engulfed with branches and foliage in years to come, so it has to be durable. Replacement or repairs might be impossible without wrecking the years of growth that it has to support. Use a wood with a good resistance to rot, or a wood which is thoroughly impregnated with preservative, which will not have an adverse effect on anything growing near it.

PERGOLA

One of the simplest structures to erect is a framework for roses or other plants to climb and cover. Such a framework on posts is called a pergola, if the assembly is more than a single-line screen. Parallel posts might be alongside a path and you could include an arch. The framework on top allows the climbers to spread and make a roof of foliage, providing an attractive arrangement, shelter from the sun, and an impenetrable shelter from rain. Besides making a floral cover over a path, a pergola can act as a shelter for seats and would make a good division in a garden, possibly between the vegetable plot and a flower and grass arrangement.

As shown in FIG. 6-1, the pergola is arranged with parallel posts, with a bay on each side of a raised part that can form an arch over a walkway. In the other direction, there could be a narrow path or the area might be taken up by the plants that produce the foliage. This design is offered as a specimen of a pergola, but you could modify the layout to suit your available space or to suit a particular garden or yard layout. It might take a few years for the foliage on a pergola to reach perfection, and you need to visualize how it will look in

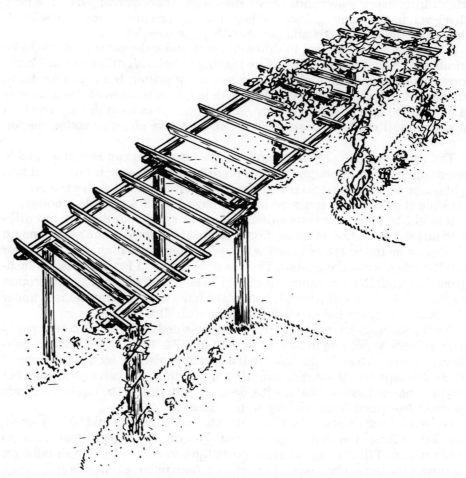

FIG. 6-1. A pergola supports foliage, so that grows to form a roof.

relation to other features. It is quite large and will form a background to a large floral display, or it might form a screen for a less attractive part of the view. It might also serve as a windbreak when the foliage is in leaf and bloom. In the winter, however, it will be more of a skeleton through which wind might blow with little hindrance, although its appearance still might be regarded as attractive.

Because of its size and the effects of wind, as well as the dead weight of what is growing on it, you should build a strong pergola. Parts cross squarely. It would be unusual to include wind bracing. Secure mounting of the posts and strong joints should provide adequate stiffness. As with other buildings, squareness is important. The ground probably will not be level. It would be a mistake to use the ground as a datum, as parts which are not vertical or horizontal will be very obvious to an observer, especially if you can see parts at different angles. Check levels and plumbs throughout the assembly. See that posts are upright, from every viewpoint. Check the levels of horizontal parts, not only with a level, but by sighting across. When you have one piece known to be level, sight across the other parts and see that they are parallel.

If the ground is not level in different areas under the pergola, it might be advisable to bring it nearer level before starting to build. A difference of 6 inches or so between one end and the other might not matter, but if you are faced with more of an incline, a large difference in height of the pergola between one end and the other will not look right, if you attempt to keep the top level. In that situation, it might be possible to get an attractive effect by taking the top up in steps.

The most vital parts of the pergola are the posts. You can mount a durable wood or other wood impregnated with preservative in the ground, but rot still might occur eventually. It might be better to keep the wood above the ground by bolting it to concrete spurs which you have buried a suitable amount.

It might be possible to buy suitable concrete spurs, but they are not difficult to make. The section of spurs should be about the same as the posts. Length will depend on the soil. Sandy soil will need longer posts than clay. Allow for about 18 inches above the ground. The example shown in FIG. 6-2A has 15 inches below ground. In most ground, it should be sufficient to compact dirt around the spur, but in very soft ground, you might have to put more concrete under and around the spurs before covering them with dirt.

A suitable mold for making a spur is a three-sided box with nails only partly driven. A block at one end makes the tapered top (FIG. 6-2B). Allow for two bolts to hold the wood. To make the holes in the concrete, place ½-inch bolts wrapped with Scotch tape into the wet cement (FIG. 6-2C), then you can knock them out after the concrete has set. Coat the inside of the box with oil or liquid detergent to prevent the wood from sticking to the concrete.

Use a sand and cement mix. If you include stones, they should be very small. After half filling the box, lay in two ¼-inch or ⅜-inch-steel rods as reinforcements. Fill the box and tamp down the concrete to remove air bubbles, then trowel the top surface level. Leave to set, then remove the nails and loosen

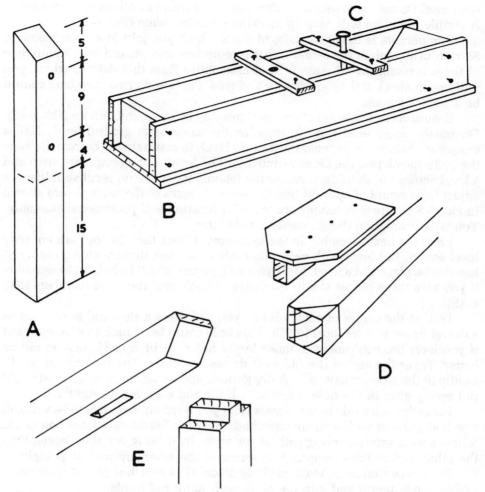

FIG. 6-2. Casting a concrete post spur (A,B,C) and woodwork details (D,E).

the wood, so the spur can be put aside to harden. Put the box back together to make the next spur.

When you bolt the posts to the spurs, make sure the bottom of the wood is a short distance above the ground. In subsequent gardening, be careful to avoid building up soil round the posts. Something should be between the wood and the concrete to insulate the wood from moisture. This insulation could be a liberal coating of a waterproof mastic or some sheet plastic, such as polyethylene. Use coach bolts with their heads at the wooden side. Grease them before driving them. Do not fully tighten until after erection, in case you must make slight adjustments if the spurs do not finish quite plumb.

Bevel the ends of all the top members. It does not matter what the exact angle is, although 45 degrees looks right. All angles, however, ought to be the same size. Do not cut to a feather edge, but finish with a small amount left square. A simple template will help in marking ends the same (FIG. 6-2D).

The pergola is based on pairs of posts which you join to a cross member. Start from an end, or the center if that is more convenient, and truly mount one of these assemblies. Measure further assemblies from this datum, which you will use to check if they are square and true. Posts and cross members should be 4 inches square.

Ensure correct location of the parts meeting by using stub tenons (FIG. 6-2E). Obviously, spacing at the top must be the same as at ground level. Nail a temporary batten across, near the ground level, to maintain the correct distance there. To check that the cross member will be level, put a temporary strip and a level across the shoulders, below the tenons. If necessary, recut shoulders or adjust the amount of spur let into the ground. Sizes of the example are shown in FIG. 6-3A. The stub tenons are there for location and provisional assembly. You will strengthen the joints with rods later.

Erect the next assembly in the same way. Check that the tops are not only level across, but level with each other when you test them with a temporary, lengthwise piece and a level. Make two long pieces. It will be helpful in assembly if you give these pieces shallow notches to hook over the cross members (FIG. 6-3B).

Drill at the center of each joint so you can drive a steel rod to be used as a dowel down into the post (6-3C). This hole could be ½ inch in diameter, but if you have the equipment to make larger holes, ⅝ inch or ¾ inch would be better. Taper the end of the rod so it drives smoothly. The rod should not fit tightly or the wood might split. A dry joint should be satisfactory, but you could put epoxy glue in the hole, especially if the rod does not fit tightly.

Make the other side in the same way. If you want the assembly to be straight, check alignment with a string stretched along the foundation line. You could follow a moderately-curving path, if you wish, by making one side shorter than the other, but in the example, it is assumed the assembly will be straight.

For the central arch, make packing pieces (FIG. 6-3D), then put lengthwise pieces above them, and join the parts with more rod joints.

What you do at the top depends on what you want to grow over the framing. You might need fairly close cross pieces for some foliage. It even might be advisable to use wire netting on a light frame. The best-looking pergola has an arrangement of cross pieces, evenly spaced, and projecting to the sides with matching ends. This arrangement is suitable for most roses, vines, and similar climbing plants.

In the example, 2-inch × 3-inch-section strips are laid across (FIG. 6-3E) and held with long nails. Place one of these pieces near each end and space the others fairly evenly between them.

If you wish to treat the entire wood with preservative or paint, choose material that does not affect plant life. It is advisable to only treat in this way well in advance of the time you expect anything to climb up and over the pergola, so any solents dry out. As plants start climbing, you might have to tie them on

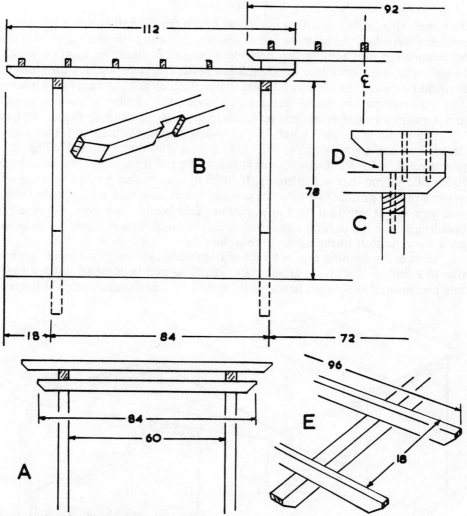

FIG. 6-3. *Suggested sizes for pergola parts.*

or provide strips of projecting wood for them to grip, then remove them after the plants have grown higher.

Materials List for Pergola

8 posts	4 × 4 ×	78
4 bearers	4 × 4 ×	86
2 beams	4 × 4 ×	94
4 beams	4 × 4 ×	114
4 packings	4 × 4 ×	18
15 tops	2 × 3 ×	98

SUMMER HOUSE

You can use a building with a sheltered porch for sunbathing or sitting out in chairs, even when the weather is not perfect, since the structure provides shelter from wind and rain. It can provide a peaceful retreat for anyone wishing to get away from activities inside the house. It might be a place for studying. It could be a play center for children, although it is not primarily a playhouse. The enclosed part of the building will provide full shelter when you need it, and it makes a place to store chairs, tables, games, equipment, or gardening tools.

The summer house shown in FIG. 6-4 has a base which is 9 feet square, divided in half by a partition with a door and windows (FIG. 6-5). The upper part is open, with sheltering lower sides, and a rail front. The door is arranged to lift off, so you can put it inside, instead of it swinging and interfering with seating on the porch. The exterior probably will look best with shiplap siding and you might use that on the partition. You could, however, cover all the building or just the partition and door with plywood. The summer house is built on a floor, which forms part of the assembly.

Most of the framing can be 2-inch-square wood, although you could increase that to 2 inch × 3 inch for greater strength. The roof is boarded, without separate purlins and is covered in the usual way. Much of the decorative appearance

FIG. 6-4. This summer house has a sheltered porch and ample inside accommodation.

144

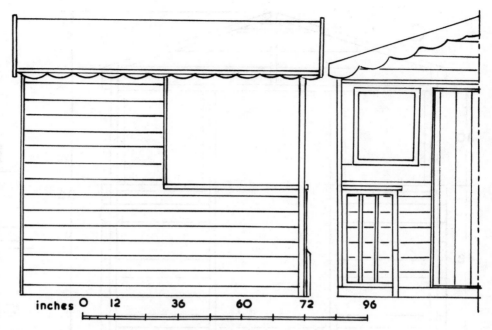

FIG. 6-5. *Two views of the summer house of the suggested size.*

comes from the bargeboards and matching eave's strips. The fence at the front has square uprights, but if you have the use of a lathe, they would look attractive if you made them as turned spindles with square ends.

Start with the floor, which should be 9 feet square. Use 1-inch boards and 2-inch × 3-inch joists at about 18-inch centers (FIG. 6-6A). Close the joist's ends with strips across (FIG. 6-6B).

Make the building to fit the floor. Let the cladding overlap the floor—either just the top boards or to the bottoms of the joists. Use the floor as a guide to sizes when making the other parts.

Make the partition (FIG. 6-6C) and use it as a height guide when making other parts. It probably will be best to make the bottom part of the frame right across at first (FIG. 6-6D), then cut out the part for the doorway when you nail or screw the partition to the floor. Halve the frame parts or use open mortise-and-tenon joints. Make the frame width to fit inside the sides when they stand on the floor (FIG. 6-6E). The side cladding should go over the edge of the floor. At the apex, allow for a 2-inch × 4-inch ridge to be slotted in (FIG. 6-6F), with a supporting rail underneath. Vary door and window sizes, if you wish.

When you have erected the building, the partition will fit between uprights on the side, with its covering overlapping, whether it is boards or plywood (FIG. 6-6G). Consequently, when you cover the partition, let the covering extend enough at the sides. At top and bottom, the covering should be level with the framing.

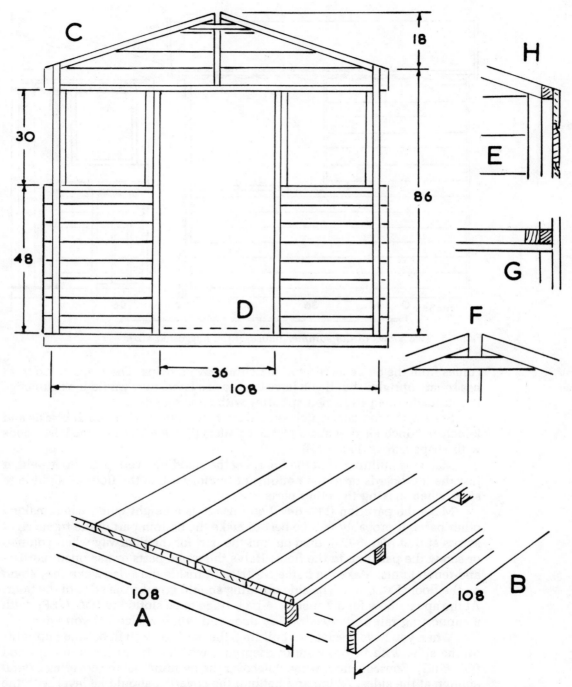

FIG. 6-6. The summer house floor (A,B), its front (C,D) and constructional details (E,F,G,H).

Make the back of the building the same as the partition, except you leave out the door and windows and extend cladding over the floor edge. This procedure means the framing could be the same as the partition, with the center and bottom rails right across. Cover in the same way as you did the partition, since there is a similar overlap at the corners, which you will cover with a filler strip between the meeting boards.

The pair of sides could have windows, but they are shown closed (FIG. 6-7A). Cladding is taken to the front edge, but you could arrange open rails, similar to the front, if you wish. Cladding should be level with the frame all around, except you should allow for going over the floor edge and for going over the ends of the covering at the front of the partition. Cover pieces will be

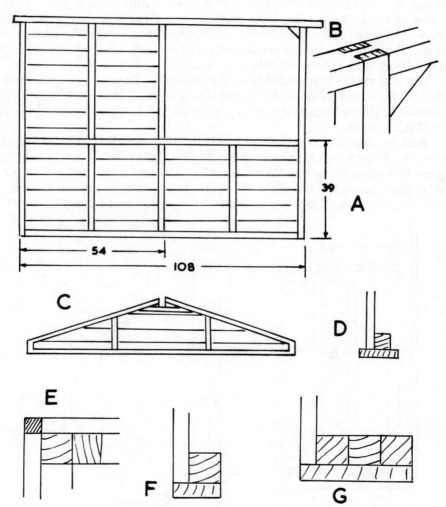

FIG. 6-7. A side of the summer house (A,B), a roof truss (C), and assembly details (D,E,F,G).

over this joint and along the top edge of the porch. Bevel the top edges of the sides to match the slope of the roof (FIG. 6-6H). The top part of the frame extends 6 inches at the front and 3 inches at the back to support the roof. This top part also might be 3 inches deep for extra stiffness and you could build small angle brackets into the front, open corners (FIG. 6-7B).

Use the top part of the partition as a guide when making the front, which fits between the side uprights (FIG. 6-7C), where you will nail and screw it. Extend its cladding over the side uprights. Slot the apex to take the ridge piece. Make its bottom edge 5 inches below the eaves. Fit a covering piece over this edge (FIG. 6-7D) and around its edges.

The rails or fence at the front are shown extending 24 inches from the sides, but you can make them any other width. This width gives a good space for moving chairs and other things in and out, as well as allowing several people to pass. Make two identical frames, with strong corner joints. Use planed wood and take the sharpness off the exposed edges. Two uprights about 1½ inch square should be enough intermediately (FIG. 6-8A).

You will mount this assembly on the edge of the floor and you will securely screw or bolt it to the floor and the side uprights. Arrange an overlapping piece to extend to the bottom of the floor (FIG. 6-8B) to stiffen the post at the open end of each piece.

Start erection of the building by bolting the two sides to the back and the partition—⅜-inch coach bolts at about 24-inch intervals should be sufficient. Square this assembly on the floor and nail the bottom edges down. Cut out the bottom piece across the doorway. Put square filler pieces in the rear corners (FIG. 6-7E). Cover exposed cladding edges at the partition and front (FIG. 6-7F). Put

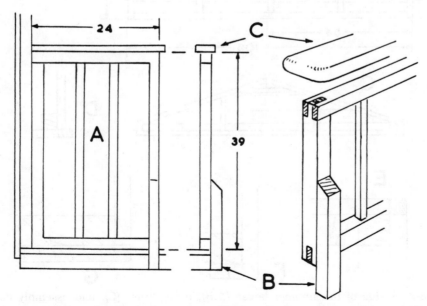

FIG. 6-8. *Sizes and details of the fence at the front of the summer house.*

148

strips on each side of the window frames so they are the same thickness as the cladding (FIG. 6-7G and 6-9A).

Fix the front rails and make an overlapping covering piece (FIG. 6-8C) with well-rounded edges. It will look best if you fix it with counterbored screws and cover them with plugs.

The two windows are shown fixed, but you could arrange for them to open, either with hinges at the top or on the outer edges. They are protected from the weather by the porch, so there is no need for a sill. Put strips all around the

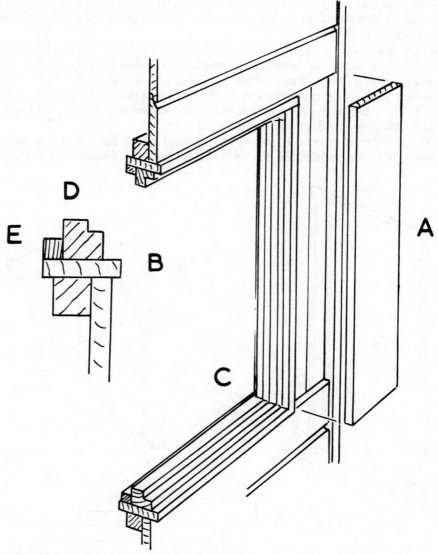

FIG. 6-9. Window construction for the summer house.

window openings (FIG. 6-9B), extending out a little, and rounding all exposed edges. Make the window frames to fit closely (FIG. 6-9C), using rabbetted strips (FIG. 6-9D). You could screw the strips directly in place, but it will probably be easier to make a good, weathertight fit with stop strips inside (FIG. 6-9E). Fit the glass with putty after you paint the woodwork.

Line the sides and top of the doorway in the same way as the window openings. Put stop pieces near the inner edges (FIG. 6-10A). Make the door (FIG. 6-10B) an easy fit in the opening. Have the edge of the bottom ledge about 2 inches from the bottom of the door. If the top ledge has only a small clearance below the top stop strip in the opening, you can fit a lock with a keyhole there, or arrange a catch which turns with a knob. Place the other ledge centrally and arrange braces both ways.

At the bottom, fit two pegs to go into holes in the floor (FIG. 6-10C). Notch over the bottom ledge and taper the extending ends slightly (FIG. 6-10D). Glue and screw these a few inches in from the sides of the door. Mark holes in the floor where you can drop the pegs in while you angle the door forward, so they hold it fairly close to the stop strips. When the top of the door is held with a lock or catch, the building will be secured.

Fit the ridge to extend 6 inches at the front and 3 inches at the back. If necessary, trim the ends of the eave's strips to the same length (FIG. 6-11A,B).

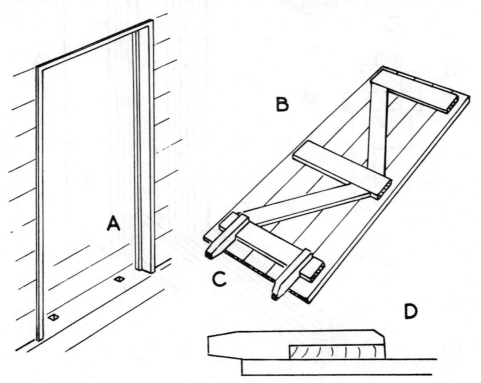

FIG. 6-10. The doorway and lift-out door for the summer house.

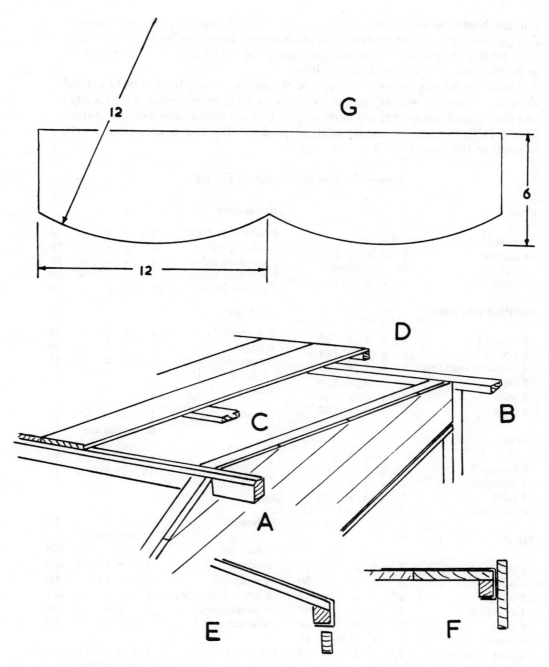

FIG. 6-11. Roof details for the summer house (A,B,C,D,E,F) and a template for marking edge decorations (G).

You can board the roof direct, using 1-inch × 6-inch boards, preferably tongue-and-grooved. If you use boards with plain edges, there can be a central batten (FIG. 6-11C) to prevent the boards from warping out of line. You do not need to fix the batten to the back or partition.

Nail the boards to the ridge and to the eaves, where they should extend about 4 inches (FIG. 6-11D). At the eaves, put a strip underneath, with its edge and the ends of the boards cut vertically, if you are adding the side decoration (FIG. 6-11E). Put similar square-edge strips down the end boards (FIG. 6-11F) to support the bargeboards.

Materials List for Summer House

Floor

7 joists	2	× 3	× 110	
14 boards	1	× 6	× 110	
	or equivalent			
2 ends	1	× 3	× 110	

Partition and back

6 uprights	2	× 2	×	88
2 uprights	2	× 2	×	24
4 window uprights	2	× 2	×	32
5 rails	2	× 2	×	110
4 rails	2	× 2	×	36
4 tops	2	× 2	×	60

Sides

8 uprights	2	× 2	×	88
2 uprights	2	× 2	×	40
4 rails	2	× 2	×	110
2 top tails	2	× 2 or 3	×	120

Front

1 rail	2	× 2	× 110	
2 rails	2	× 2	× 60	
3 uprights	2	× 2	× 24	
4 uprights	2	× 2	× 42	
4 rails	2	× 2	× 28	
4 posts	1¼	× 1¼	× 42	
2 post supports	2	× 2	× 24	
2 rail tops	1¼	× 3	× 28	

Edge covers

2 side-edge tops	1¼	× 4	×	54
2 side uprights	1	× 4	×	70
2 window sides	1	× 7	×	30
2 window sides	1	× 5	×	30
2 corner fillers	1	× 1	×	88

Windows

8 surrounds	1	× 4	×	28
8 stops	1	× 1	×	28
8 frames	2	× 2	×	28

Door

6 boards	1	× 6	×	80
	or equivalent			
3 ledges	1	× 6	×	36
2 braces	1	× 6	×	36
2 pegs	1½	× 1½	×	12

Roof

32 boards		× 6	×	60
	or equivalent			
2 battens	1	× 3	×	108
2 edges	2	× 1¼	×	120
4 ends	1¼	× 1¼	×	60
2 edge decorations	1	× 3	×	120
4 bargeboards	1	× 6	×	66
8 battens	¼	× 1	×	60

Cladding   1- × -6 shiplap boards or equivalent

Carry roof covering over from eaves to eaves and turn under for nailing. Turn under at the ends. Allow ample overlap where there are any joints and make joints in the direction that will let water run away from them. You could add capping strips, but they probably will not be necessary on this small roof. Nail battens down the slope at each side at about 18-inch intervals.

If the decoration on the lower edges of the bargeboards and eave's boards is to look right, the curves should be uniform. Make a template of at least two curves, using scrap plywood or hardboard (FIG. 6-11G). Use this template to mark all the shaped edges and to check them after shaping. Nail the boards to the roof to complete construction.

FOLDING-DOOR SUN LOUNGE

When the weather is bright and warm, you might wish to take the sun directly, but in cooler conditions, you might enjoy it better through glass. Even on a dull day, it might be pleasant to sit behind an expanse of glass while sheltered from the wind.

This sun lounge is a complete building (FIG. 6-12), but it is arranged so you can open most of the high side to give you shelter while letting the sun in. If all or some of the doors are closed, you are sheltered more, although the glass lets the sun shine through.

Of course, if the object of the building is to let in sunlight, it has to face the sun, so you have to choose a location facing south. You might have uses for such a building if it does not face south, but if the main purpose is to let in the sun, you need to face it south where the sun's rays are not shaded by

FIG. 6-12. The sun lounge has a front made of doors which fold back.

trees or buildings. Allow for the lower arc of the sun in winter, if you want to get the most from the sun lounge in cooler weather.

The building is large enough for several other uses. You could enjoy a hobby there. Children might use it as a play room. At a sports field or recreational area, you could use it for storage and for viewing events, or as a judge's enclosure. Spectators could find shelter inside while participants brave inclement weather or just a passing storm. For year-round use, line the walls and, preferably, insulate them. Double glazing would help, but if you plan to use the building after dark, heavy drapes would more than serve the same purpose as secondary glass.

As shown in FIG. 6-13, the building is 6 feet × 12 feet and the doors and back wall are 6 feet 6 inches high. The four front doors are each 30 inches wide, so they open to 10 feet. A single door is in the back, and there can be windows in the ends. It might be possible to buy suitable, standard doors. If so, you might have to modify some dimensions to suit them. The making of the doors is included in the following instructions.

The drawings show a covering of vertical, tongue-and-groove boards, but you could use horizontal shiplap boards or sheets of plywood or other covering material. The framing suits upright boarding. If you use horizontal boarding, you might need more uprights for its support.

The roof is given a solid, square-edged appearance and there are no bargeboards or fascias. Several roof coverings are possible. Treated felt or similar material could be laid over boards. It would be possible to use wood or composite shingles on boards or plywood. There is a solid-wood edging all around.

You could lay a wooden floor on joists over a concrete slab, arranging the structure so the walls stand on the floor and the sheathing continues over the floor. An alternative is to mount the building directly on concrete, stones, or bricks, and fit the floor inside after erection. If you wish to make the floor first, follow the instructions used in earlier examples. The following instructions are for fitting a floor inside the building. This type floor is slightly shallower, which might be an advantage if you do not want much of a step up from the surrounding surface.

You might make the main structure of softwood, but a durable hardwood would be better for the doors, which have to be strong enough to withstand rough use. There would be a risk of broken glass if they were weak enough to flex when moved violently, possibly by the wind. If it is possible to get tongue-and-groove boards in the same wood, you would obtain an attractive effect by giving it a natural look with an oil or varnish finish.

You could erect the building before you make the doors, but as their sizes are important to the rest of the assembly, you might prefer to make them first, then you can allow for any slight differences in the doorway with less trouble than if you had to alter the sizes of the doors. The rear door, if fitted, might be almost identical to the others, or you can panel it fully with wood, instead of having the glass panels.

Prepare the parts for all the doors together and cut and fit joints at the same time, so they finish identical. Use planed wood, which will be about ¼ inch

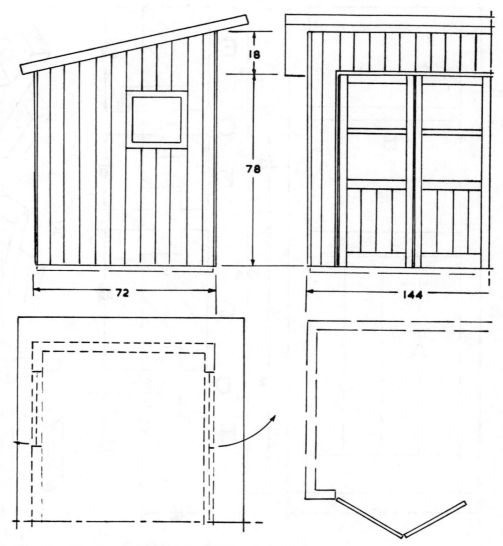

FIG. 6-13. Sizes and layout of the folding door sun lounge.

under the nominal size, so 2 inch × 3 inch for the top and sides will actually be 1¾ inch × 2¾ inch. Make the door with the lower part filled with tongue-and-groove boards (FIG. 6-14A). The glass panels have their bar slightly above halfway, which looks better than dividing the space equally (FIG. 6-14B).

Rabbets will have to be put in the upper parts for the glass and grooves in the lower part for the boarding. If the rabbets are made 1 inch wide (FIG. 6-14C), the tenons might be ½ inch wide. You can continue this design into the lower part, where the grooves are ½ inch wide (FIG. 6-14D) and the mortises and tenons will fit into them. Prepare the top of the door with a rabbet right through

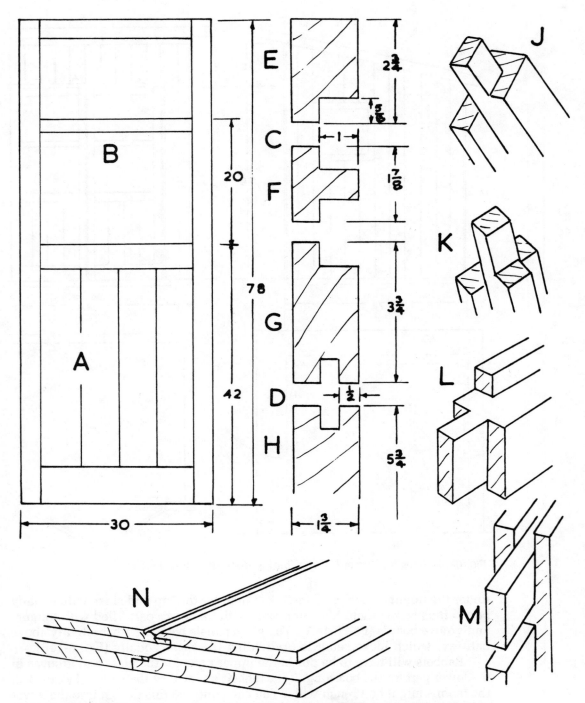

FIG. 6-14. Make the sun-lounge doors first. Here we show them with traditional joints.

(FIG. 6-14E). Make the central-glazing bar with similar rabbets on both sides (FIG. 6-14F). The bar between the glass and wooden panels needs a rabbet on the top and a groove underneath (FIG. 6-14G). For the bottom of the door, cut a groove right across (FIG. 6-14H). Prepare the door sides in pairs with rabbets down to the dividing bar and grooves below that.

Have the door sides too long until after you have cut mortises and glued in the tenons. This procedure prevents the end grain from breaking out and protects the doors while they are handled, if the extensions are not cut off until you are about to fit the doors. Tenons need not go through the sides, but should be about 1½ inches deep.

Mark all the crosswise parts together, so they are the same length between shoulders. Cut back the tenon widths at the top, so there is solid wood left in the sides and set back the shoulders to suit the rabbets (FIG. 6-14J). Cut mortises in the sides to suit. Cut the glazing bar similarly (FIG. 6-14K).

At the center bar, cut the shoulders the same length and also the tenons between the bottoms of the rabbets and the bottoms of the grooves. Notch the back to fit over the rabbetted part of each side (FIG. 6-14L). At the bottom, cut back the tenon width in the same way as the top. Because of the depth, divide the tenon into two, with a ½-inch gap (FIG. 6-14M).

If you can get tongue-and-groove boards only ½ inch thick, they could go directly into the grooves, but the boards are more likely to be thicker than that. If so, cut down the edges to fit in the grooves (FIG. 6-14N). Do not force them too tight in the width, as there should be a little allowance for expansion and contraction. Lengthwise, they should be a tight fit.

Be careful to check squareness as you assemble the doors. Check the shape of each door on those you previously assembled. See that they remain flat, as a twist in any door will affect the fit of the whole set. If you clamp tightly as you assemble and use a waterproof glue, the door joints should be strong, but you could drill through the centers of the tenons and glue ½-inch dowels through the joints.

Although you might not be painting or varnishing yet, you should apply putty over paint. It is worthwhile painting in the rabbets now, so it is dry when you start glazing.

The pair of sides are straightforward frames (FIG. 6-15A). Use 2-inch × 3-inch wood with the 2 inch direction towards the skin. Corner joints might be any of the usual type, although open mortise-and-tenon joints (FIG. 6-15B) should make the strongest corners. If you will be covering with vertical boards, there is no need for intermediate uprights. If you will be using horizontal-shiplap boards, continue the window upright nearer the center to the full depth. Check squareness and see that the opposite ends match.

Cover with upright tongue-and-groove boards, making them level with the edges all around. When you have erected the building, cover the meeting corners with a filler strip (FIG. 6-15C).

Trim the boards level with the frame around the window. At the top, fit a covering piece with a taper and groove to shed rainwater (FIG. 6-15D). At the bottom, arrange a similar piece (FIG. 6-15E). Notch both covering pieces and

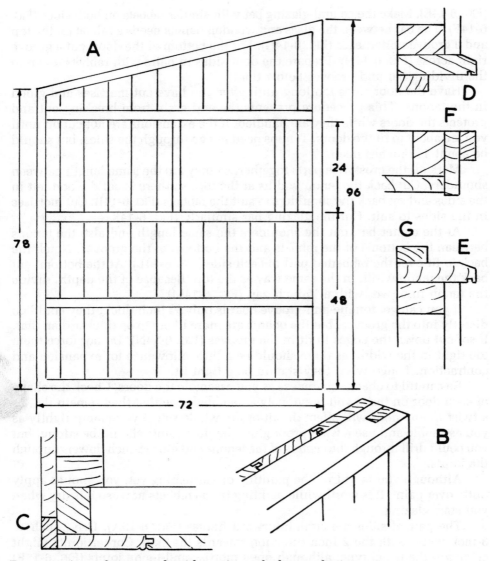

FIG. 6-15. A sun lounge end, with sections of edges and joints.

continue them a short distance over adjoining boards. Cover the vertical edges with narrower strips (FIG. 6-15F). Put stop pieces all around inside the frame (FIG. 6-15G).

Frame the back (FIG. 6-16A) with 2-inch × 3-inch strips with their 2-inch face towards the skin, except for the top piece. Face the top piece the other way, to allow enough wood for the bevel to match the roof (FIG. 6-16B). Position the rear door, if you want one, in a position to suit your needs. It cannot be quite as high as the front doors.

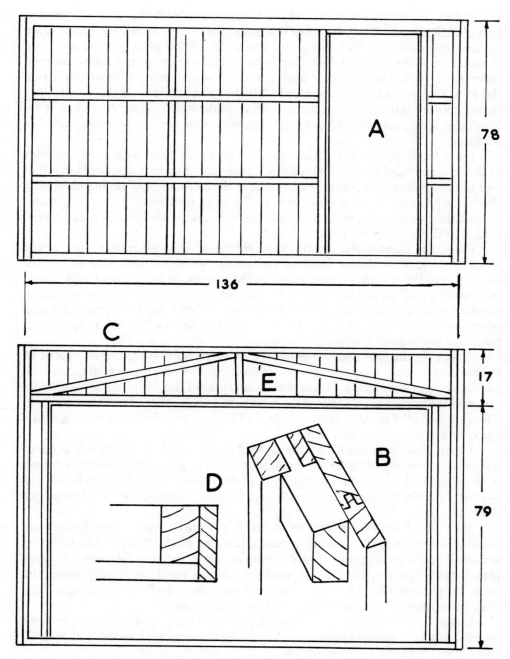

FIG. 6-16. Details of the back and front of the sun lounge.

When you cover the back, cut the boards level with the frame at top and bottom, but allow extra at the sides to overlap the ends (FIG. 6-15C). The overlap should be an amount that will allow you to fit a filler piece during erection. If you cover with horizontal-shiplap boards, include more uprights at 30-inch intervals or less. The length of back and front shown allows for fitting inside the ends to give an overall length of 12 feet. You can increase this length, but do not shorten it if you are to include four front doors which are 30 inches wide. You must have some of the front boards on each side of the doorway to provide stiffness.

Make the front (FIG. 6-16C) in a similar way to the back, with a strip-on edge at the top, bevelled to match the slope of the roof. Measure the doors together and allow for the cover pieces (FIG. 6-16D) at the sides and top when positioning the frame parts for the doorway. To keep the long, horizontal parts straight and to hold the frame in shape, join a short, central upright and two diagonal braces (FIG. 6-16E) to the other parts.

Cover the front in the same way as the back, with extending parts at the sides. Trim level with the doorway and cover the edges. Each side piece will have to take the weight of two doors on the hinges, so fasten the cover pieces with waterproof glues and screws. When you mount the doors, use screws in the hinges long enough to penetrate the wood behind the door frame. Fit stop pieces at top and sides that will hold the front surfaces of the doors level with the front surfaces of the wall.

Assemble the four walls together with ½-inch bolts at the corners about 18 inches apart. Check squareness by measuring and comparing diagonals. If necessary, put temporary diagonals across to hold the building square while you add the roof, which is based on 2-inch × 4-inch rafters from front to back. The finished roof is intended to project 12 inches at the front, 6 inches at the back, and 9 inches at each side.

Fit the first two rafters over the ends of the building. Other rafters at about 24-inch intervals should be adequate (FIG. 6-17A). Join the ends of the rafters with pieces that project far enough to hold the rafters at the side overhang (FIG. 6-17B). Provide further support with short pieces between the end rafters (FIG. 6-17C).

You could cover the roof with thick plywood, but 6-inch boards are suggested (FIG. 6-17D). Shingles, plastic tiles, or any of the flexible covering materials could go over these boards. After covering, edge all around with strips (FIG. 6-17E). The strips at the front and sides could stand up slightly, but at the back, keep the top edge low, so it does not stop the runoff of water (FIG. 6-17F). You could mount a gutter there, leading to a downpipe.

Cover the undersides of the overhanging rafters. You could cover with boards, but it is probably easier to use ½-inch-exterior plywood (FIG. 6-17G) fitted closely to the wall boarding.

The windows in the end might be fixed or opening, hinged at the top or side. You can putty glass for a fixed window directly into the stop strips, but it is better to make separate frames, as described for earlier building (FIG. 5-25).

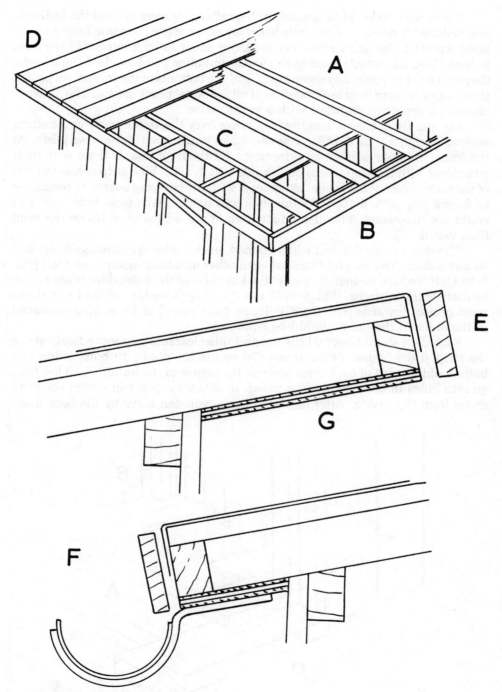

FIG. 6-17. Roof construction of the sun lounge.

If you wish to build in a wooden floor after you have erected the building and fastened it down to a concrete base, lay down joists the same height as the bottom parts of the wall frames (FIG. 6-18A) at about 18-inch intervals from front to back. You can attach them to the walls, but after you have laid the boards, they will hold in place. Lay these joists, and the bottoms of the frames, on plastic sheeting or be sure they are coated well with a waterproof mastic. They always should be protected from rot with a preservative.

Lay the floor boards lengthwise, going over the frame parts and cutting around uprights (FIG. 6-18B). You can make joints lengthwise over joists. At the front, the floor will act as a doorstop. Fit the floor boards there with their edge level with the stop pieces on the side (FIG. 6-18C). Fit a strip across in front of the bottom part of the frame, either level with the covering boards or projecting to form a step (FIG. 6-18D). As this area will have to take wear from feet, you could use hardwood. You also might wish to put a hardwood lip on the front floor board.

The doors close flat, but when opened, each center door swings in against its outer door. The two fold against each other and back against the wall (FIG. 6-19A). If they are to open this way, the knuckles of the outer door hinges must be clear of the surfaces, (FIG. 6-19B) and the hinge knuckles of the inner doors must project inwards (FIG. 6-19C). Three 4-inch steel or brass hinges, placed in the sides of the door, should be suitable.

Each door should have a little top and side clearance and move freely above the strip at the bottom of the frame. To secure the doors, fit bolts at top and bottom of the edge of each door towards the center of the building, so they can go into holes in the surrounding wood. In this way, you can secure the front doors from the inside. After fastening them, you can leave by the back door.

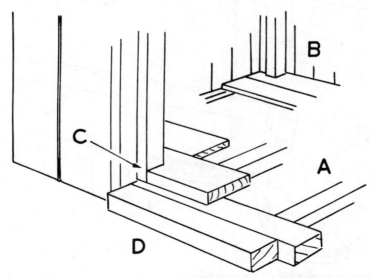

FIG. 6-18. Details of the floor and front corner of the sun lounge.

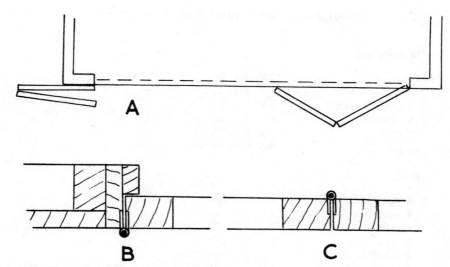

FIG. 6-19. *How you fold and hinge the doors of the sun lounge.*

You do not need a fastener between the inner doors and you do not need to fit handles as you operate the doors by pushing.

Fit the back door in a similar way, but provide it with a lock operable from either side and a knob or handle.

HEXAGONAL GAZEBO

Although most buildings have square corners, they can be other shapes. You might want to fit a building into an awkwardly-shaped plot of ground. This requirement would need to be individually designing. Buildings might be made of special shapes for the sake of their appearance. A round building is difficult to make in wood, but a multi-sided one need not be much more difficult to construct than a square one. Making a building with an odd number of sides is an interesting geometric problem. An eight-sided building, not necessarily a regular octagon, involves angles of 45 degrees, but you will be dealing with twice as many sides as the more usual square-cornered building. If you have six sides, you will not need to make so many walls and roof sections. Some advantages exist in giving the building a regular hexagon as a floor plan. One advantage is the attractive appearance. Another is the ease with which you can set it out.

The sides of a regular hexagon are the same length as the radius of the circle on which you base it, so the lines from the corners to the center divide the area into equilateral triangles (FIG. 6-20A). As all the angles in an equilateral triangle are 60 degrees, this is convenient. You can set many saws and planers accurately to this angle.

Materials List for Folding-Door Sun Lounge

Doors (five)

10 sides	2 × 3 × 80
5 tops	2 × 3 × 30
5 glazing bars	2 × 2 × 30
5 center rails	2 × 4 × 30
5 bottoms	2 × 6 × 30
20 panels	1 × 6 × 40 tongue-and-groove

Ends

4 uprights	2 × 3 × 100
4 uprights	2 × 3 × 80
8 rails	2 × 3 × 74
2 tops	2 × 3 × 78
4 window sides	2 × 3 × 26
4 window linings	1 × 4 × 26
4 window frames	2 × 2 × 26

Back

5 uprights	2 × 3 × 80
2 rails	2 × 3 × 140
2 rails	2 × 3 × 100
2 door linings	1 × 4 × 78
1 door lining	1 × 4 × 32
2 door stops	1 × 2 × 78
1 door stop	1 × 2 × 32

Front

2 uprights	2 × 3 × 100
2 uprights	2 × 3 × 80
3 rails	2 × 3 × 140
1 upright	2 × 3 × 18
2 diagonals	2 × 3 × 74
2 door linings	1 × 4 × 80
1 door lining	1 × 4 × 122

Roof

10 rafters	2 × 4 × 96
2 rafter ends	2 × 4 × 168
6 spacers	2 × 4 × 12
2 edges	1 × 4 × 96
2 edges	1 × 4 × 168
16 boards	1 × 6 × 168 or plywood ½ to fit under

Floor

8 joists	2 × 2 × 72
12 boards	1 × 6 × 144 or equivalent
1 step	2 × 4 × 122

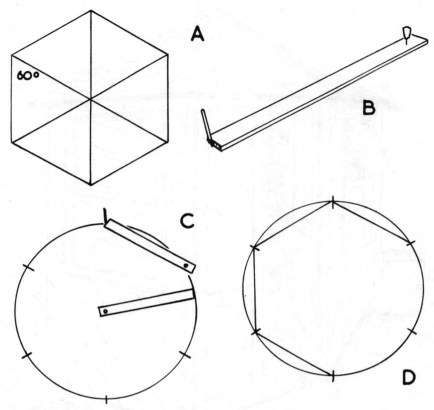

FIG. 6-20. The method of marking out a large hexagon.

To set out a large hexagon, make an improvised compass with a strip of wood, placing an awl through at half the distance you want the shape to be across the points (FIG. 6-20B). With a pencil against the end, draw a circle. Now move your "compass" so you can step off the radius around the circumference (FIG. 6-20C). Join the points you have marked to make your regular hexagon (FIG. 6-20D). You can make a hexagon using the size across the flats, but it is easier to work using the distance across the points.

The gazebo shown in FIG. 6-21 and 6-22 is based on a hexagon which is 9 feet across the points. It is 7 feet from the floor to the eaves. One side has glazed double door. The sides next to it have windows. The other three sides are boarded solidly.

The covering is shown as tongue-and-groove boards laid diagonally. There could be horizontal-shiplap boards or you could have tongue-and-groove boards vertically. A painted plywood skin might suit some locations. Shingles are suggested for the roof, but you can use other coverings.

To mark out the floor shape, put down sufficient floor boards with their undersides upwards. With an improvised compass, obtain the positions of the points and join them to get the hexagonal floor shape (FIG. 6-23A). It is

FIG. 6-21. This hexagonal gazebo is 9 feet across and has two windows and double doors.

convenient to have the floor boards parallel with two sides. On this floor shape, lay out the floor beams, which you might halve at the corners and other joints (FIG. 6-23B).

Assemble the beams together, then turn the floor over and nail down the floor boards. Trim the edges level. This design should make a strong, flat assembly on which you can base the building (FIG. 6-23C). If it is to be mounted on concrete, you might wish to raise it with blocks at the corners, so there is ventilation below.

You can mount the walls on the floor in two ways. Keep the wall framing level with the edges of the floor, then continue the cladding down to cover the

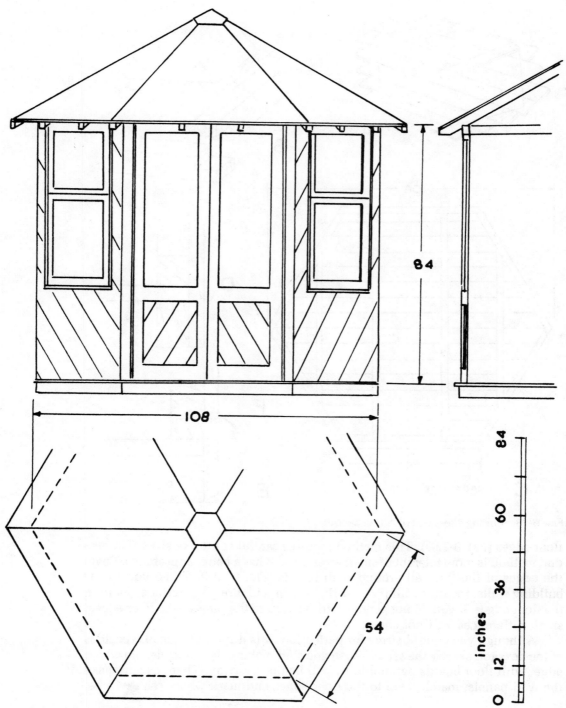

FIG. 6-22. The sizes and shape of the hexagonal gazebo.

167

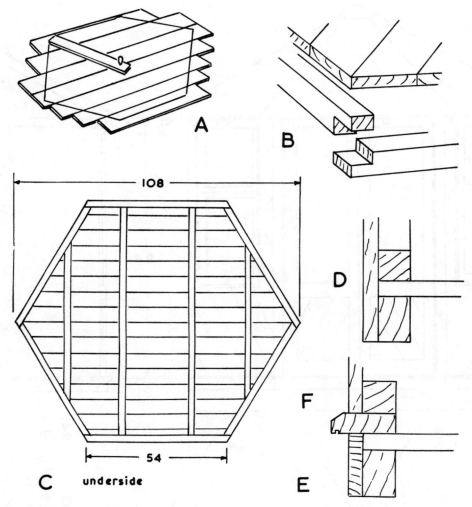

FIG. 6-23. *Making the floor of the gazebo.*

floor edges (FIG. 6-23D). This method requires careful control of sizes. The second method is more tolerant of small errors, and it has a better appearance. Cover the edges of the floor all around with boards (FIG. 6-23E). When you fit the building walls, include a sill (FIG. 6-23F), which will direct rainwater away from the floor edges. If you do not get the wall sizes quite the same as the floor edges, small differences will not show.

Although you should strive for perfection, it is difficult to get all six sides of the hexagon exactly the same. When you start making the walls, decide which edge, with floor boards parallel to it, will be the doorway. Then, as you make the wall panels, match them to the floor edges and mark where each will be.

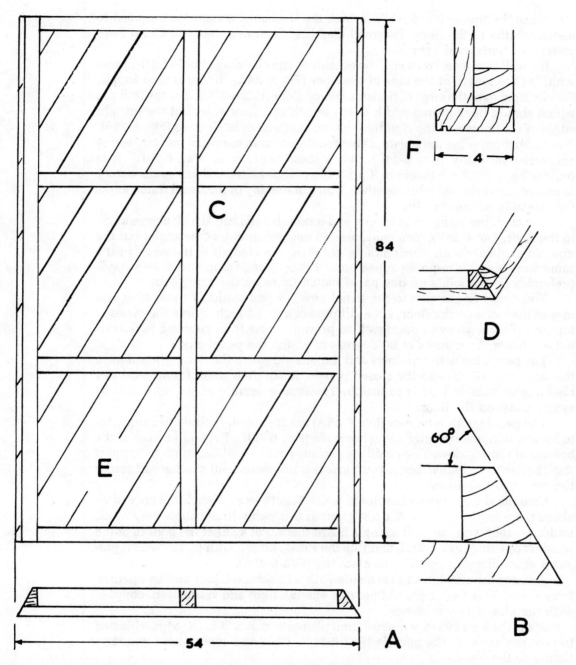

FIG. 6-24. One solid side of the gazebo, showing diagonal boards and sections at corners and bottom.

Make the three closed walls, which are the same, except for the need to match widths to the floor. Diagonal cladding is shown, but you could have vertical or horizontal boards.

If a wall is to stand on a sill, the outside of the cladding should be the same width as the length of the side of the floor (FIG. 6-24A). If you intend to take the cladding over the edge of the unbordered floor (FIG. 6-23D), it is the framing which should be the same width as the length of a floor side. Cut the upright edges of the frame and the cladding, when you fit it, to 60 degrees (FIG. 6-24B). Top and bottom edges are square. Other framing divides the panel into two across and three vertically (FIG. 6-24C). When these walls meet (FIG. 6-24D), it is impossible to put bolts through. It probably will be easier and more satisfactory to nail or screw the uprights together. You can use any of the usual frame joints for assembly of these walls.

The cladding is shown as tongue-and-groove boards laid at a 60 degree angle to the floor (FIG. 6-24E). You could use 45 degrees or any other angle, but the upward angle gives an illusion of height. You can clad all of the walls in the same way, but you improve appearance if you arrange slopes alternate ways, preferably so the joints on one panel match those on their neighbor.

You could fit the sills to the panel now, or you could fit them after you mount the walls on the floor. Let a sill project about 1 inch. Slope the exposed top and plow a groove underneath to prevent water from running back (FIG. 6-24F). Miter the corners at 60 degrees to match the panel sides.

The two sides with windows and the one side with the doors should have the same overall sizes as the closed panels. Make their outer frames and any cladding or outside boards to match. The final assembly of six sides will be symmetrical on the floor.

The panels with windows (FIG. 6-25A) are identical. Include two uprights to leave a window opening 40 inches wide (FIG. 6-25B). Put a rail across the bottom of the window. The cladding probably will be stiff enough to support itself below the window, but if you think it is necessary, put another rail across the center of this space.

Clad the sides to match the closed walls. Short pieces of cladding are shown above the window space (FIG. 6-25C), but as this part will not show very much under the roof, you may put a single board there. Put a sill at the bottom of the window opening (FIG. 6-25D) and line the sides and top with pieces which project a short distance outside the cladding (FIG. 6-25E).

Arrange the windows in each side with a lower fixed part and an opening top section. This design should provide enough light and ventilation, coupled with the glazed pair of doors.

Make the lower fixed windows from planed 2-inch × 3-inch strips rabbetted to take the glass on the outside (FIG. 6-25F). Glue and screw these windows tightly to the frames.

For the opening windows, put stop strips around the opening (FIG. 6-25G). Make the window with a 2-inch-square top and sides, and a 3-inch-deep bottom (FIG. 6-25H). Hinge the window at the top, and fit a strut and catch to the lower edge.

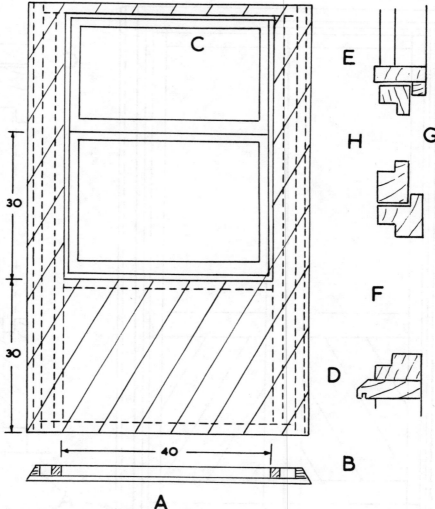

FIG. 6-25. A gazebo side with a window.

When you make the frame for the pair of doors, make a temporary bottom rail to keep the frame in shape until you erect it. When you assemble the six sides on the sill and floor, take that rail out, so the doors fit over the sill and swing inwards (FIG. 6-26A).

You will not have space to fit diagonal cladding on each side of the doorway, but you should fill in with vertical pieces (FIG. 6-26B) and fill the top to the same thickness. Line the sides and top (FIG. 6-26C). Make the two doors the same and in the same way as described for the previous building (FIG. 6-14). There is no intermediate glazing bar and you should arrange the bottom panels with boards diagonally. Reduce the thickness of the boards to fit in the rabbets the same way as the earlier door.

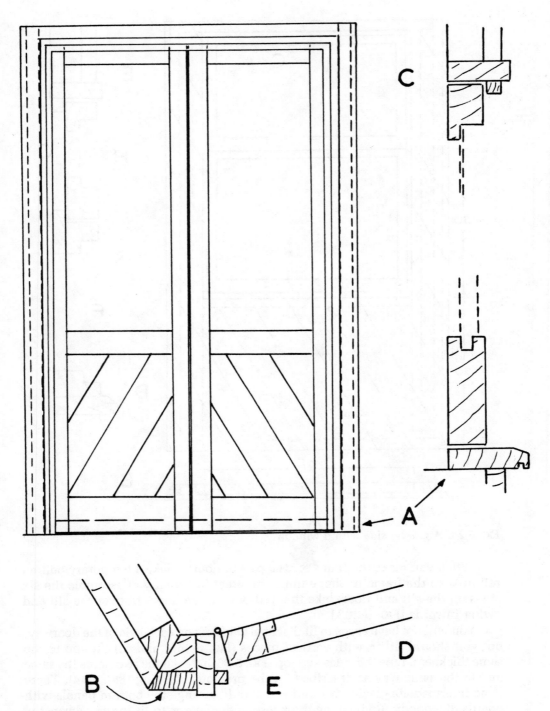

FIG. 6-26. The gazebo side containing the doors.

Hinge the doors to swing inwards (FIG. 6-26D) and put strips outside at top and sides (FIG. 6-26E). Three 4-inch hinges should be satisfactory. One door might have bolts up and down and the other door might have a lock to it.

Although it is possible to prefabricate the floor and the six sides almost completely before erection, it is advisable to work on the roof in position. Build the walls on the floor before starting on the roof. See that there is no twist. The floor will keep the bottom in shape, but check diagonals at the top. If there are any errors, put temporary struts between offending corners and leave them in place until the roof is far enough advanced to hold the walls in the correct shape.

At the top, fit 2-inch- × -3-inch strips laid flat to serve as wall plates when you build on the roof (FIG. 6-27A). To help tie the walls together, halve and screw the corner joints in this assembly.

The roof slopes at 30 degrees from the corners of the building (FIG. 6-27B). As the distance is less across the sides, it is slightly steeper on the surfaces. Although you can do a certain amount of preparatory work, it is better to do much of the fitting of parts as you progress.

A pair of rafters across opposite corners are the key parts in making the roof (FIG. 6-27C). Cut their meeting ends at 60 degrees and link them with a piece underneath (FIG. 6-27D). Notch over the wall plates and trim the ends to shape (FIG. 6-27E). The roof covering has to rest on the rafters, so bevel their top edges to suit. The angle shown in FIG. 6-27F is approximate and you should test this when you have temporarily assembled all the rafters, by laying a straight piece of wood across. It does not matter if you do not achieve perfection, but it is easier to cover the roof neatly if parts cross reasonably flat.

Make the rafters to the other corners in the same way, but allow for the thickness of the main rafters at the top (FIG. 6-27G). When you are satisfied with the top angles and assembly so far, nail the rafters to the wall plates and to each other. For most roof coverings, it is advisable to add more rafters spacing them equally along each side (FIG. 6-27H). Their tops are flat. Cut their lower ends in line with those of the corner rafters. This step completes the skeleton of the roof. Check flatness of the surfaces in all directions.

Whatever the method of covering the roof, it will help to first cover the rafters with exterior plywood. This plywood gives a smooth base and acts as a lining for the roof. Use ½-inch plywood (FIG. 6-28A). Where plywood joints come on the six main rafters, cover with sheet plastic or other waterproof, flexible material, with a good overlap, and glue down or make sure it is held in place by covering it later. Cover any joints in the plywood in the same way.

Arrange battens to suit the shingles, if that is the type of covering you prefer (FIG. 6-28B). As shown in FIG. 6-28C, the shingles are 16 inches long with a 4-inch exposure (FIG. 6-28C), so the battens are 4-inch centers. Work from the eaves upwards. Where the shingles meet over the main rafters, miter them. You could interleave more plastic or bituminous sheeting under adjoining shingles, for further waterproofing.

At the apex, the shingles will reduce to quite narrow triangles. Take them as close as you can to the top, then use a hexagon of lead or copper sheeting about 12 inches across, to cover the meeting of the six sets of shingles. Finish it to the roof shape and nail it down closely.

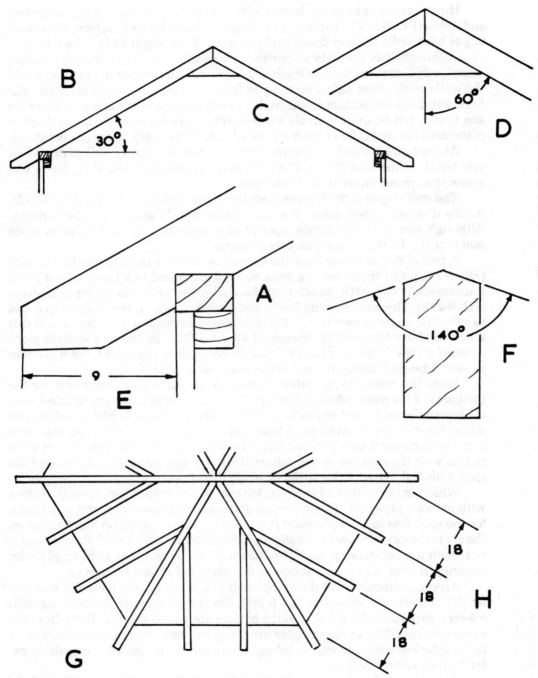

FIG. 6-27. Roof construction of the gazebo.

Materials List for Hexagonal Gazebo

Floor

6 joists	2	× 3	×	58
2 joists	2	× 3	×	112
2 joists	2	× 3	×	72
6 edges	1	× 4	×	58

Covering   1 × 6 or equivalent

Sills   1¼ × 4 × 60

Closed walls

9 uprights	2	× 2	×	86
12 rails	2	× 2	×	56
Covering	1- × -6 tongue-and-groove or equivalent			

Window walls

8 uprights	2	× 2	×	86
4 rails	2	× 2	×	56
2 rails	2	× 2	×	46
2 window linings	1	× 3½	×	42
4 window linings	1	× 3½	×	54
2 window stops	1	× 1	×	42
4 window stops	1	× 1	×	26
4 window frames	2	× 2	×	42
2 window frames	2	× 2½	×	42
2 window frames	2	× 3	×	42
4 window sides	2	× 2	×	24
4 window sides	2	× 3	×	30

Door wall

2 uprights	2	× 2	×	86
2 rails	2	× 2	×	56
1 door frame	1	× 3½	×	52
2 door frames	1	× 3½	×	84
1 door stop	1	× 1	×	52
2 door stops	1	× 1	×	84
2 side pieces	1	× 3½	×	86
4 door sides	2	× 3	×	84
2 door tops	2	× 3	×	22
2 door rails	2	× 6	×	22
2 door rails	2	× 4	×	22
Panels	1- × -6 tongue-and-groove or equivalent			

Roof

6 rafters	2	× 4	×	78
12 rafters	2	× 4	×	48
Covering	½ plywood, 1- × -2 battens and 16-inch shingles			

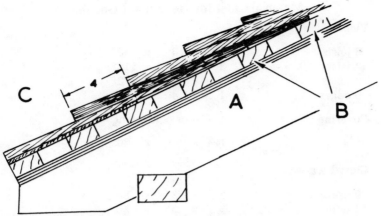

FIG. 6-28. A roof section showing shingles over plywood.

A gap will be between the wall plates and the underside of the roof. This gap might be left as ventilation, but if you prefer it closed, fit pieces between the rafters.

The entrance is a few inches above the surrounding ground. Fit a step under the sill, and make it the full width of the doors. It will be convenient and improve the appearance of the gazebo.

ARBOR

If you make a pergola structure with a seat, it becomes an arbor. Foliage grows over it and around it and makes it a shelter. The foliage might be quite dense, except at the front, or it might just form a roof. Roses particularly are associated with an arbor, but you can use any type of climbing plant. You could build a pergola and place a seat under it, but it is better to build the seat in. As this is a permanent structure and since it is exposed to all kinds of weather, you must make the seat of wood. Any softening must be with portable cushions. This fact does not mean the seat cannot be at a comfortable angle, for use without cushions on occasions.

The arbor shown in FIG. 6-29 has inverted V legs supporting a flat top similar to a pergola. At each end, a strut parallel with a leg slopes up to give additional support to the top. This strut sets the angle of the seat back which has vertical slats. You can make the bottom of the seat solid or slatted. The legs go into the ground and crosspieces prevent them from sinking too far. The suggested sizes (FIG. 6-30A) are for an arbor 6 feet high and about 7 feet long, but you can alter these to suit your needs or available space.

You can use softwood treated with preservative or a more durable hardwood. Remember that once you erect the arbor and foliage is growing over it, you cannot do much to treat or repair it. Its original construction must be strong enough to have an expected life of many years. Bolts ought to be galvanized to minimize rust. Any glue should be a waterproof type and screws should be plated

FIG. 6-29. Make an arbor like a pergola so foliage can form a roof over a seat.

or should be made of a noncorrosive metal. The main parts are 2 inch × 3 inch or a 4-inch section.

Start by setting out an end. If you want to set tools, the angles are 15 degrees. From the ground line, draw a centerline square to it, then 72 inches up, mark the apex of a triangle with a 39-inch spread at the base (FIG. 6-31A). Draw the top across (FIG. 6-31B) and mark the widths of the wood. Draw the seat support across the legs (FIG. 6-31C). The seat top is symmetrical about the centerline of this seat support and the strut slopes up from the back of it, parallel with the front leg (FIG. 6-31D). This layout gives you all the shapes and sizes you need to start construction.

Notch the tops to take the beams (FIG. 6-31E). Let the legs meet on the beam and drill through for bolts. Spread the bottoms and secure them with the ground strips (FIG. 6-31F). On this framework, mark where the seat bearers come. Put these pieces across and add the long struts (FIG. 6-31G) marking where they cross and where you want to drill for bolts or where you want to cut joints.

On the seat bearers, mark and cut the notches for the lengthwise seat supports, not more than 1½ inches deep (FIG. 6-31H). The strut joins by halving. This halving is best cut with a dovetail shape (FIG. 6-31J). Cut the notches to take the lengthwise back supports (FIG. 6-31K). At the top, halve the strut into

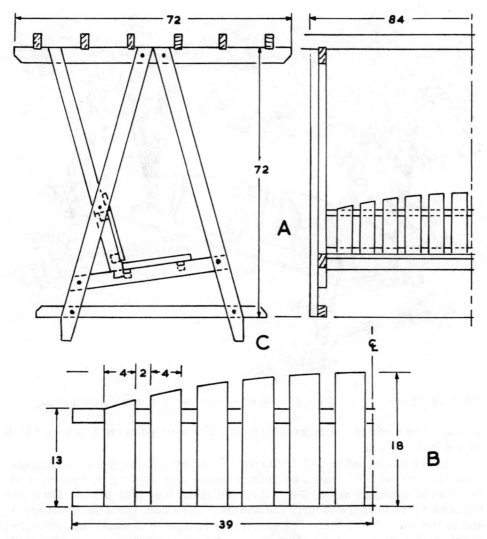

FIG. 6-30. Sizes of the arbor.

the top piece. Assemble the pair of ends, with bolts where parts overlap and glue and screws at the joints.

The seat back has vertical slats with a curve cut on their top edge (FIG. 6-30B). You might prefer some other shape. Make the two, 2-inch-square rails, notching ends to fit the notches in the struts. Have the back slats too long at first. Fit them temporarily to the rails. Bend a batten over them and draw a curve on their tops. Remove the slats to cut their curved tops and round all exposed edges. Glue and screw them in place.

Make the seat supports to the same length as the back supports. Put pieces between them—if you divide the length into three, that should be sufficient (FIG.

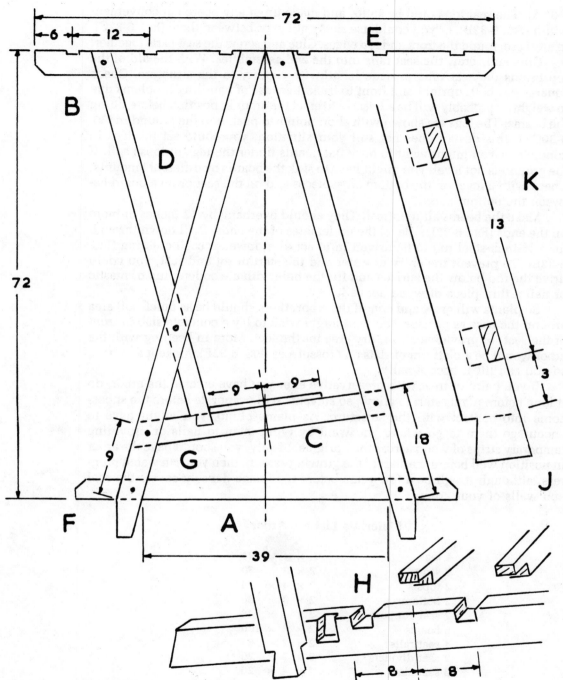

FIG. 6-31. Detail of one end of the arbor.

6-32A). The seat top could be solid, and made up of any boards of convenient width (FIG. 6-32B), or you could use slats with gaps between them (FIG. 6-32C). In any case, round the front and top edges. Glue and screw the seat parts together.

Glue and screw the seat rails into the end assemblies. With the aid of the top beams the seat rails provide lengthwise rigidity to the structure. Check squareness, both upright and front to back. Because of handling problems due to weight, it probably will be advisable to erect the arbor in position before fitting the beams. The legs are shown with short points to push into the ground in FIG. 6-30C. If this design does not suit your situation, you could set the legs in concrete or you might want to bury flat boards under the legs in loose soil. If the ground is not level, you might have to sink the boards by different amounts. Check with a level on the bottom crosspieces and on the seat or on a board between the bottom parts.

Make the beams all identical. They should overhang by 12 inches or more on the ends (FIG. 6-32D). Bevel the undersides of the ends. Drill down through for a ½-inch-steel rod to be driven in to act as a dowel at each crossing (FIG. 6-32E). To prevent the entry of water and the start of rot and rust, you could drive the rod below the surface and fill the hole with a wooden plug or mastic or nail a thin piece of wood above it.

So plants will grow and engulf the arbor, there should be as much soil area around the base as possible, but you might wish to lay a concrete slab in front of the seat, to provide a clean, dry area for the feet. More in keeping with the arbor would be a platform of slats on crosspieces (FIG. 6-32F). Make it as a unit, so you can lift it occasionally.

If you paint or treat with preservative after you have erected the arbor, do it long before plants start to climb, so solvents will evaporate before the shoots come into contact with the structure. As plants climb, you might have to encourage them to go where you want by tying them to nails or providing temporary strips of wood across the uprights. Ideally, you should have the arbor in position well before the start of the growing season, then you can watch progress, although it will be a few years before the folage densely covers the roof and walls of your arbor.

Materials List for Arbor

4 legs	2	× 4	×	84
2 struts	2	× 3	×	70
2 tops	2	× 4	×	74
2 bottoms	2	× 3	×	66
2 seat supports	2	× 4	×	48
6 beams	2	× 4	×	120
4 seat rails	2	× 2	×	84
2 seat dividers	2	× 2	×	20
12 seat slats	1	× 4	×	20
2 seat boards	1	× 6	×	84 or slats
4 platform strips	1	× 4	×	74
6 platform supports	2	× 2	×	22

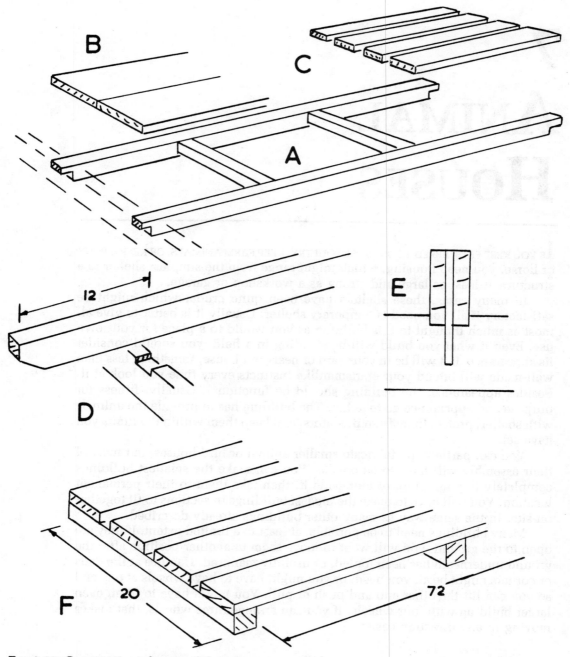

FIG. 6-32. Construction of parts of the arbor.

7

ANIMAL HOUSES

IF YOU KEEP PETS, BREED BIRDS, RAISE POULTRY, KEEP FARM ANIMALS, OR KEEP A PONY or horse, you need housing, which might range from the simplest shelter to a structure at least as large and strong as a workshop or garage.

In many cases, these shelters have been quite crude, which might be satisfactory if all you need is temporary shelter. Usually it is better to give almost as much thought to this building as you would to a place for your own use. Even if what you build will be standing in a field, you should consider its appearance. If it will be in your yard or near your house, something less than well-made will offend your craftsmanlike instincts every time you look at it. Besides appearance, the building should be functional. Usually, fitness for purpose and appearance go together. The building has to provide the animals with shelter, protect them from predators, and keep them within the limits you have set.

You can partially prefabricate smaller animal or bird houses, but most of their assembly will have to be on-site. You can make the smallest buildings completely in your shop, or alongside it, then take them to their permanent location. You will have to make the biggest buildings in sections to fit together on-site, in the same way as many other buildings already described.

Many buildings need to be portable, at least to a limited extent. If they are open to the ground, you will want to move them to another position after the ground underneath has been fouled, or usefully manured. Handles at the ends or corners might be all you need, or you might have to have wheels at one end so you can lift the other end and push or pull. You might have to fit an even larger building with four wheels. If you can arrange three wheels, that makes moving in any direction easier.

ARK

A small shelter, light enough for two people to carry, is useful for small birds and animals, such as ducks, which might wander freely during the day, but which need a closed shelter at night. An *ark* is a simple, V-roofed building. You can vary sizes to suit the animals. Pigs will need a much bigger ark than ducks or other domestic birds. The ark shown in FIG. 7-1 measures 60 inches in all directions and will suit a small flock of ducks or hens. It does not have a bottom and you can move it to a new position every day, if necessary. One end has a lifting door. This shelter needs to have an opening big enough for the birds or animals to pass easily, but it should not be any larger. It is shown in FIG. 7-2A about 18 inches high and 16 inches wide.

Make the door end first. Halve the bottom corner joints and the end of the bar above the opening with its door uprights. At the top, cut down for a 1-inch-thick capping and notch to make a tenon on a ridge piece (FIG. 7-3A). To avoid the complication of a three-way, sloping-bottom corner joint, notch the roof framing for lengthwise pieces above the corner joint (FIG. 7-2B and 7-3B).

Cover the framing with horizontal boarding. These boards could be plywood, shiplap boards, tongue-and-groove-boards, or just plain boards. The door might be a piece of ½-inch, exterior-grade plywood (FIG. 7-2C). Make guides with

FIG. 7-1. You can move this light ark for small animals or birds easily.

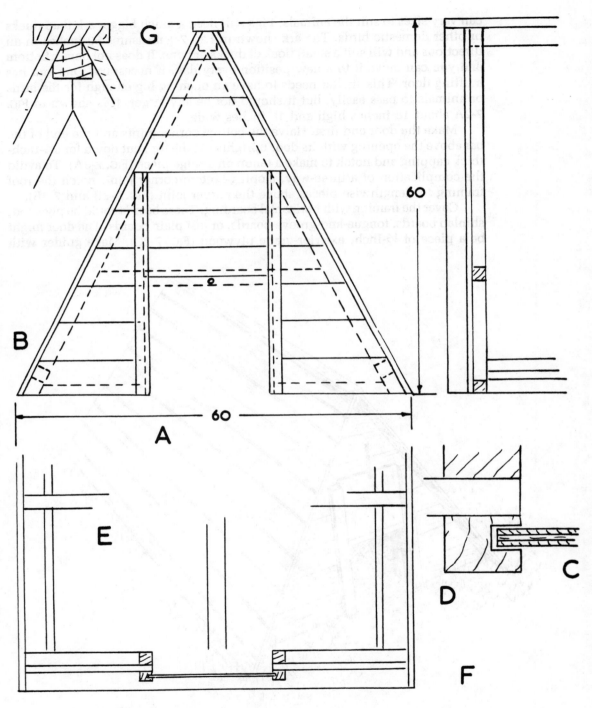

FIG. 7-2. Sizes and sections of the ark.

grooves (FIG. 7-2D). These guides will keep the door ¼ inch forward of the covering boards and allow it to slide very easily. With a hole at the top of the door, you can use a cord or wire to a hook or nail to hold the door up. You cannot lift the door out, so it is unlikely that you will lose it during the day. You can remove it if you lift the end of the ark off the floor.

Make the opposite end to match. It could also have a door, but it is shown closed in FIG. 7-3C. Take the bottom rail right across and include a central post, if your covering needs stiffening. In both ends, finish the covering boards level with the framing all around, including the doorway edges.

Make the ridge piece (FIG. 7-3D) with its edges matching the slopes of the sides and with tenons to fit in the slots in the tops of the ends. Make the two bottom side rails with their ends halved to fit in the notches in the ark ends (FIG. 7-3E).

Whether you should include any other framing or not depends on the stiffness of the covering boards and the length of the ark. For a length of 60 inches, it probably will be advisable to include central rafters (FIG. 7-2E and 7-3F) between the bottom rails and the ridge piece.

The roof covering might be exterior-grade plywood, simple, overlapping weather boarding (FIG. 7-3G), or shiplap boards (FIG. 7-3H). Let the boards extend about 3 inches past the ark ends (FIG. 7-2F). At the ridge, flatten the edges of the boards and nail on a capping (FIG. 7-2G).

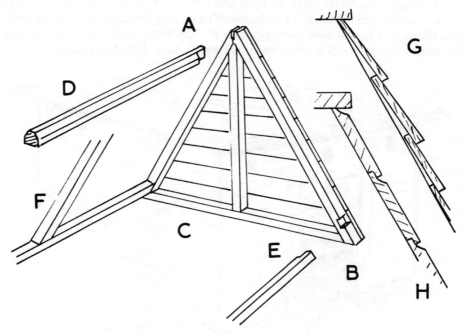

FIG. 7-3. Constructional details of the ark.

Materials List for Ark

6 rafters	2 × 2 × 70
1 bottom	2 × 2 × 6
2 bottoms	2 × 2 × 24
1 rail	2 × 2 × 48
1 post	2 × 2 × 60
2 door sides	2 × 2 × 40
2 side rails	2 × 2 × 60
1 ridge	2 × 2 × 60
1 capping	2 × 4 × 64
1 door	18 × -18- × -½ plywood

Covering 1-×-6 boards, plain, shiplap, or tongue-and-groove

MINI HEN HOUSE

If you want to keep just a few laying hens, you do not need a large house for them. This poultry house is intended for four hens or maybe six bantams. You could use it for ducks, but you would have to fit it with a solid floor instead of the slats. The mini hen house shown in FIG. 7-4 and 7-5 is without a run, but if you want to stop your little flock from wandering too far, there is a suitable run described next.

Construction might be quite light and a skin of plywood on 1-inch × 2-inch strips is suggested. There is a pop hole with a sliding door. One perch is provided and there is a single nest box opposite it. The roof hinges open for access to the entire house, which you can move easily to a different position.

FIG. 7-4. A mini hen house made of plywood and suitable for a few laying hens.

186

Make the front (FIG. 7-5A) with the framing having its 2-inch side against the plywood for top and sides, but having the 1-inch side towards the plywood across the bottom. The hole is 11 inches wide and high. Make the door and its guides in the same way as for the ark (FIG. 7-2). Frame the back in the same way as the front, but make it 24 inches high. Drill 2-inch ventilation holes high in each part.

Make the pair of sides (FIG. 7-5B). Use the framing strips in the same manner as the back and front, but at the corners, allow for the plywood to overlap (FIG. 7-5C). Bevel the top edge of the front to match the slope of the sides. There is no need to bevel the top edge of the back. Notch the corner strips over each other at the bottom, when you nail the walls together.

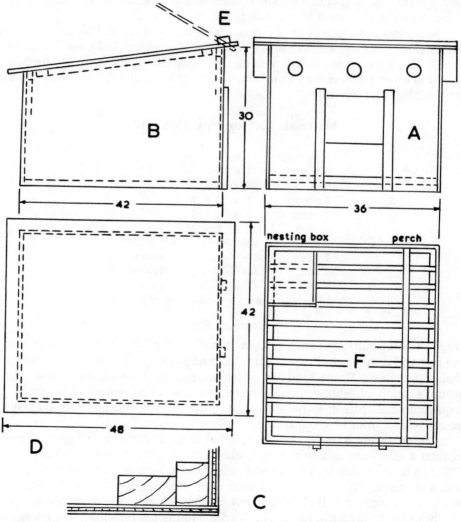

FIG. 7-5. *Sizes and internal arrangements of the mini hen house.*

The roof is a single piece of plywood overhanging 3 inches all around (FIG. 7-5D). Frame around the underside to fit very loosely inside the walls—this prevents roof warping since you do not make a close fit in the walls. At the front, where you are to place the hinges, put a strip across above the plywood (FIG. 7-5E) to take the screws. Two, 3-inch hinges should be adequate. You could have a fastener at the back, but its weight will hold the roof closed.

Put supports for the perch (FIG. 7-6A) at both ends, about 6 inches from the bottom and side. The perch is a square section, but take sharpness off the edges.

Make the slatted floor (FIG. 7-5F) of 1-inch-square strips with rounded edges, nailing them 3 inches apart (FIG. 7-6B) to the bottom strips on the sides. If the house is for ducks, put in a plywood floor in place of the slats, or on top of the slats, if you might want to use the house for hens later.

Make the nesting box of plywood, with framing strips holding it into the corner (FIG. 7-6C). The sizes shown should suit one bird, but check your bird's needs. The only access to the nesting box is by lifting the roof. No feeding or drinking arrangements are in the house, since it is assumed it is only to be used for sleeping and nesting.

Materials List for Mini Hen House

7 pieces	1 × 2 × 38
6 pieces	1 × 2 × 44
6 pieces	1 × 2 × 32
4 pieces	1 × 2 × 26
1 perch	1 × 1 × 44
9 slats	1 × 1 × 38
2 sides	30- × -42- × -$\frac{3}{8}$ plywood
1 front	30- × -38- × -$\frac{3}{8}$ plywood
1 back	26- × -38- × -$\frac{3}{8}$ plywood
1 roof	42- × -50- × -$\frac{3}{8}$ plywood
Nest box and door	12- × -36- × -$\frac{3}{8}$ plywood

Chicken Run

With a small poultry house or ark you can allow the birds to roam freely during the day and only be shut up at night. Occasions exist when you need to restrict their movement. Some kinds of chickens require limited movement. Small animals, such as rabbits, need a restricted run. For the occasions when you require a small run, it is helpful to have one which you can put against any house with a door at ground level.

The run shown in FIG. 7-7 is intended to go directly on the ground and fit against a pop hole in a hen house, allowing you to operate the sliding door. This run has two side frames, an end, and three top sections, one of which might hinge upwards (FIG. 7-8A). Fit one-inch wire mesh inside all parts. Construction is mainly 1-inch × 3-inch strips, but the bottom rails are 4 inches deep.

Start by making the two sides. Various joints are possible, but we suggest you glue two ½-inch dowels each place (FIG. 7-8B). Match the two assemblies

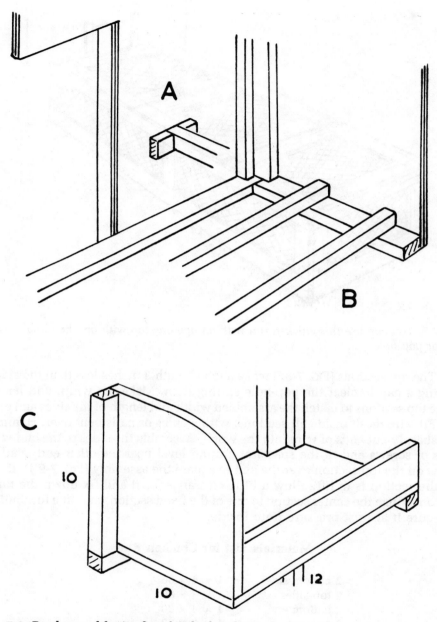

FIG. 7-6. Perches and laying box for the hen house.

(FIG. 7-9A). Make the end the same height to fit between the sides (FIG. 7-8C and 7-9B). To hold the other ends of the sides the same distance apart at ground level, make a thin, plywood spacer about 9 inches wide (FIG. 7-8D and 7-9C), which you will fix between the sides, 3 inches from the end (FIG. 7-8E) when you assemble the run.

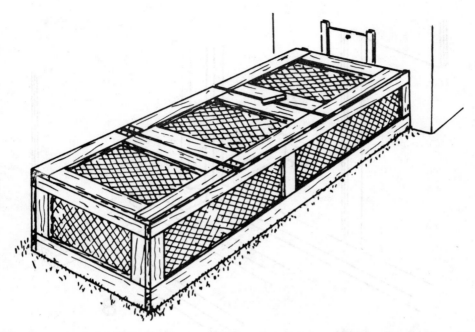

FIG. 7-7. *You can use this chicken run with an opening top with any hen house with a floor pop hole.*

The top sections (FIG. 7-8F) make a total length 3 inches less than the sides, leaving a gap to clear the pop-hole sliding door. Make the width and length of the top sections to match the assembled width and length of the sides and end.

Fit wire mesh inside all sections, with staples or nails bent over. Hammer any sharply-cut ends of wire into the wood. Assemble the sides to the end with nails or screws and fit the spacer at ground level near the other end. Nail or screw on the top section over the end to square the assembly (FIG. 7-9D). If the middle section is to lift, allow a little clearance for it and fix down the third section. Hinge the center section to one of the fixed sections and fit a turnbutton to secure it against pressure from inside.

Materials List for Chicken Run

2 bottom sides	1 × 4 × 86
2 top sides	1 × 3 × 86
2 bottom ends	1 × 4 × 30
2 top ends	1 × 3 × 30
8 uprights	1 × 3 × 22
12 tops	1 × 3 × 30
1 spacer	9- × -30- × -¼ plywood

190

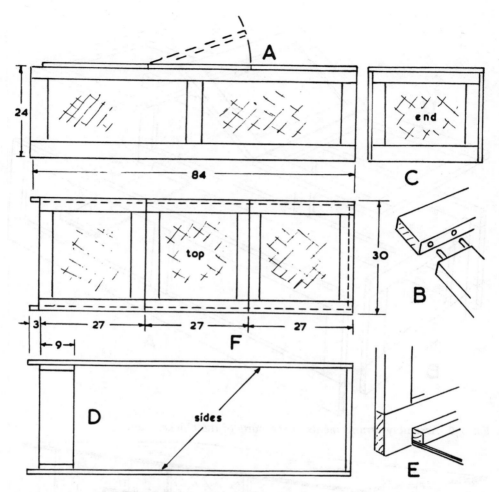

FIG. 7-8. *Sizes and details of the chicken run.*

AVIARY HEN HOUSE

If you want to keep a few laying hens where they are unable to roam freely, you have to provide a run as well as a house. This run should be more lofty than a low run which might be adequate for temporary use or for chickens.

This aviary hen house (FIG. 7-10 and 7-11A) is intended to be as compact as possible for six or eight hens. Arrange extra run space by having the floor of the house above a part of the run, with a ramp from the doorway. If you want to close the door, hinge the ramp up and hold it with a turnbutton. There is a nest box long enough for at least three hens and doors to the run and the house.

Construction is light, with plywood, which need not be more than ¼-inch thick, on wood which is mostly 1½-inch-square section. Fix wire mesh inside all the run sections. The complete roof is plywood, although in a suitable climate, you could have a solid roof on the house and more wire mesh over the run. In

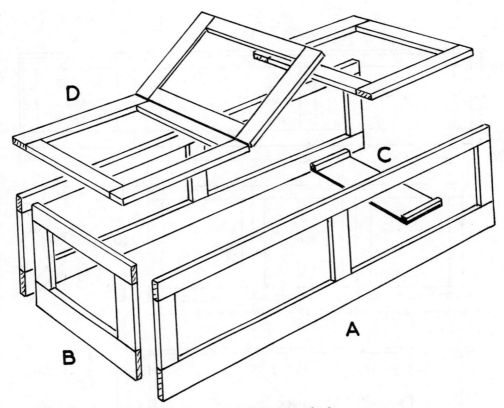

FIG. 7-9. Assembly arrangements of the parts of the chicken run.

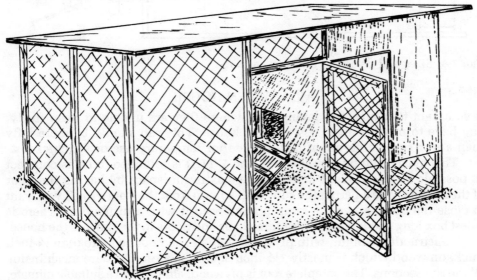

FIG. 7-10. The aviary hen house is a self-contained unit.

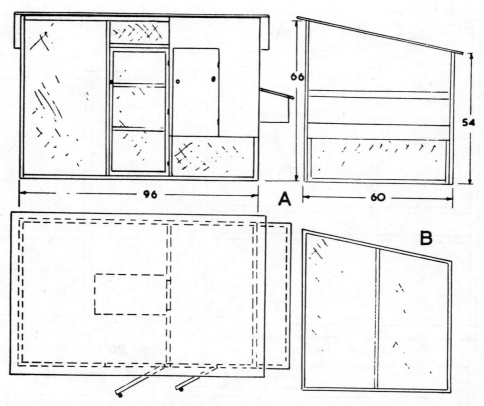

FIG. 7-11. Sizes and layout of the aviary hen house.

an area where birds need protection from a strong prevailing wind, you can replace some of the wire mesh suggested with more plywood panels. Because of the fairly light construction, strengthen many corner joints with small, metal triangles nailing them on one or both sides. Parts which might seem weak when you prefabricate them will gain strength when you assemble them to other parts, which provide mutual support. Although you can make most sections away from the site, you should nail or screw them together since this is not intended to be a portable building.

You must match the three crosswise sections in their general outlines. At the house end, cover the frame with plywood, except for the gap for the nesting box and the wire below (FIG. 7-12A). At the other end is a matching open frame with a central upright, covered with wire mesh (FIG. 7-11B). Inside is a partition without the bottom crossbar (FIG. 7-13A). The back is a simple frame with plywood behind the house and wire mesh elsewhere. The front (FIG. 7-12C) has both doors.

You can make the aviary hen house either way around and you can alter sizes. Decide how you want the assembly and make the house end first (FIG. 7-12A). Use it as a pattern for the outline of the opposite end (FIG. 7-11B), which

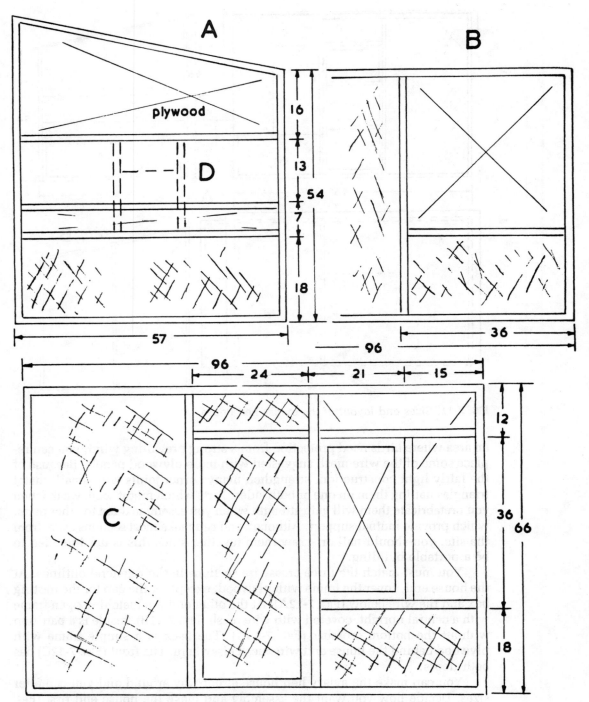

FIG. 7-12. Outside parts of the aviary hen house.

you will cover entirely with wire. A central upright might be enough for stiffness, but you might wish to add a central, horizontal rail. For the partition (FIG. 7-13A), make the house end to match the other end. Make it in the same way, but leave out the lower nest box rail and put in two uprights at the sides of a 12-inch-square opening (FIG. 7-12D).

Make the back with pieces to match the house framing on the other parts (FIG. 7-12B). Include one or two uprights in the open part to stiffen it and provide attachment for the wire mesh. Cover the house back with plywood.

When you make the front (FIG. 7-12C), bevel the top pieces to match the slope of the roof. It is not as important to bevel the top of the back section. Fit framing to suit the house parts. Framing for the door openings should provide enough stiffness for this house without adding any additional upright in the run section.

The house door (FIG. 7-14A) is framed plywood. Arrange it to fit in the opening. Put a short or full length stop inside. The hinges will hold the other edge, and a wooden turnbutton will keep the door closed.

Stiffen the run door with two strips across (FIG. 7-14B). Place three hinges and a stop and a turnbutton on the door similar to the house door.

Make the ramp (FIG. 7-13B) from a board which is 4 inches wider than the doorway, and which is long enough to slope somewhere between 30 degrees and 45 degrees to the ground. Put slats across for the hens to grip. Cut the top slat to an angle which fits against the partition, so you can screw hinges outside

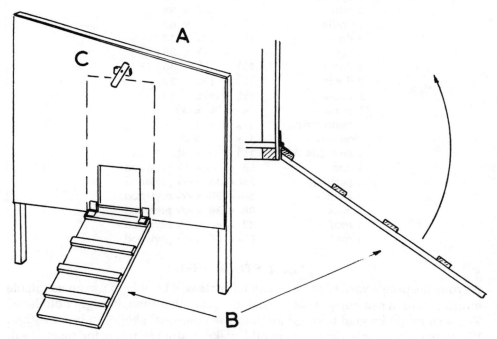

FIG. 7-13. House side and ramp for the aviary hen house.

the opening, then you can swing the ramp up and hold it with a block and turnbutton above (FIG. 7-13C).

The nesting box is held in place by its ends and bottom fitting into the frame (FIG. 7-14C). Frame the plywood with strips. For the lid, put a solid wooden strip against the wall (FIG. 7-14D) and hinge a framed plywood lid to it (FIG. 7-14E). You can fit all of these pieces to the end before you assemble the building.

Attach wire mesh to the insides of the frames—1-inch or 1¼-inch mesh should be satisfactory. Use staples or nails with their heads bent over. Hammer any cut ends of wire into the wood.

Level the ground where the aviary hen house is to be placed, leveling it completely or where the edges will rest, so you can erect it without distortion. Screws are preferable to nails, as you can join the parts tightly and there is no risk of damage from hard hammering. Join the back and front to the partition, then the house end, and the other end. Put in a sheet-plywood floor to the house part. This floor will square the whole assembly before you fasten it down.

If you use ½-inch plywood for the roof, it probably will be stiff enough without framing. Allow it to overhang at least 3 inches all around. A joint can be over the partition. You can stiffen the edges with strips underneath, which would be advisable if you want to cover the wood (FIG. 7-14F).

Materials List for Aviary Hen House

7 uprights	1½ × 1½ × 68
7 uprights	1½ × 1½ × 56
4 rails	1½ × 1½ × 98
14 rails	1½ × 1½ × 65
2 doors	1½ × 1½ × 56
2 doors	1½ × 1½ × 26
2 doors	1½ × 1½ × 36
2 doors	1½ × 1½ × 36
2 doors	1½ × 1½ × 23
1 ramp	16 × 26 × ¾
5 ramp strips	1 × 2 × 17
Nest box	1 × 1 × 240
1 nest box lid	1½ × 3 × 38
1 end	58- × -30- × -¼ plywood
1 partition	58- × -30- × -¼ plywood
1 front	50- × -38- × -¼ plywood
1 back	38- × -38- × -¼ plywood
1 roof	42- × -66- × -½ plywood
1 roof	66- × -66- × -½ plywood

SMALL STOCK SHED

Anyone keeping a variety of birds and animals will be able to use an adaptable building which can house a pig or goat, or, at another time, it might hold hens. You also might be glad to use if for your own personal shelter in bad weather. When not used for animals, it would make a storage place for tools, feed, fertilizer, and many other things needed about a small farm.

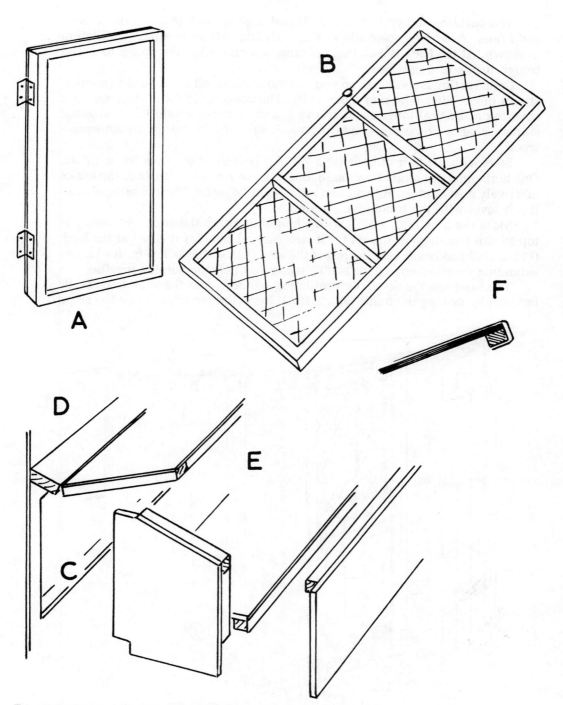

FIG. 7-14. Doors and nesting box details for the aviary hen house.

The building shown in FIG. 7-15 is just high enough to stand in. It has a solid floor. A wire mesh window with a sliding shutter to adjust ventilation is shown. At the back is a slotted opening, which could give access to a feed trough inside or a nest box.

Construction is sectional, so you can move the small stock shed to another site in pieces after releasing a few bolts. The drawing in FIG. 7-16A shows a covering of vertical tongue-and-groove boards, but you could use horizontal shiplap boards or exterior-grade plywood. Framing is mostly 2-inch-square wood.

Start with the pair of ends (FIG. 7-17A). Halve or tenon the frame joints. One horizontal rail might be enough, but if the skin has to withstand the antics of a lively goat or other animal, two rails might be better. The cladding should finish level with the framing all around.

Make the back (FIG. 7-17B) with a height to match the ends and bevel the top to suit the slope of the roof. Fit two parallel rails for the gap at the back (FIG. 7-17C) and board over the rest of the area. At the side, allow for the boards extending over the ends (FIG. 7-16B) when you bolt the corners together.

The front has the door, the ventilating window, and the slide to cover it, but start by making the frame (FIG. 7-18A). Bevel the top edge to suit the slope

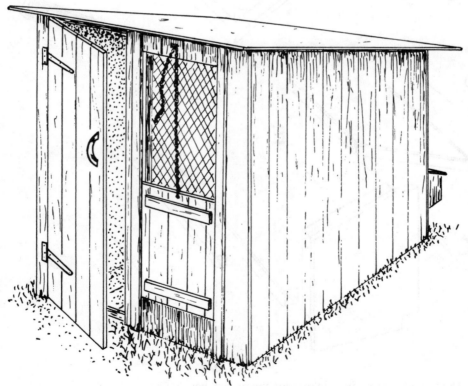

FIG. 7-15. You can adapt a small stock shed to many purposes.

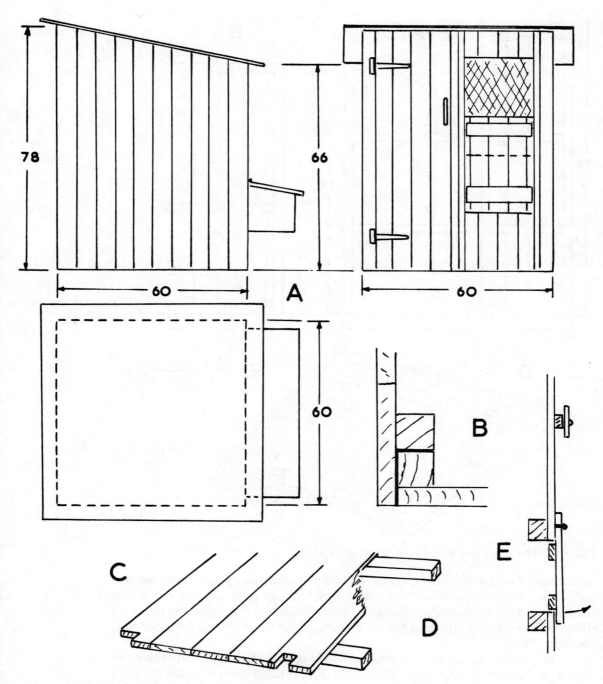

FIG. 7-16. Sizes and details of the small stock shed.

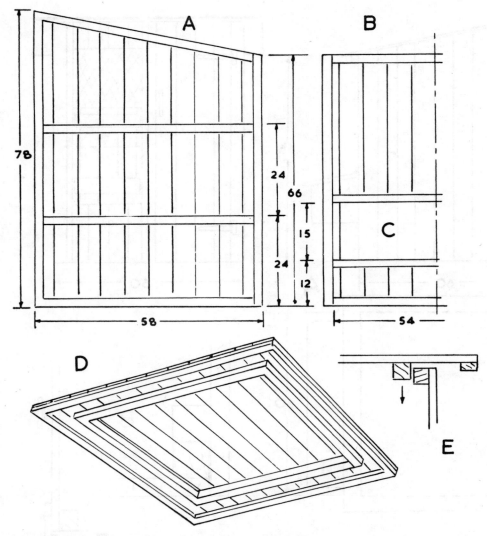

FIG. 7-17. The sides and roof of the small stock shed.

of the roof. For the window opening, put strips inside the uprights between the rails (FIG. 7-18B). Cover with boards up to halfway over the upright at the doorway (FIG. 7-18C). At the other side, finish the board level with the doorway (FIG. 7-18D). Do not fill in at top or bottom of the doorway, as the door will overlap there.

The guides for the shutter are the full height of the front. Cut rabbets to allow the shutter to slide easily (FIG. 7-18C,E). Fit them parallel at the opening width. Make the shutter from tongue-and-groove boards with ledges across (FIG. 7-18F), keeping the ends of the ledges clear of the guides. Fix a length of chain to the center of the top of the front, then you can hook the chain on to hold

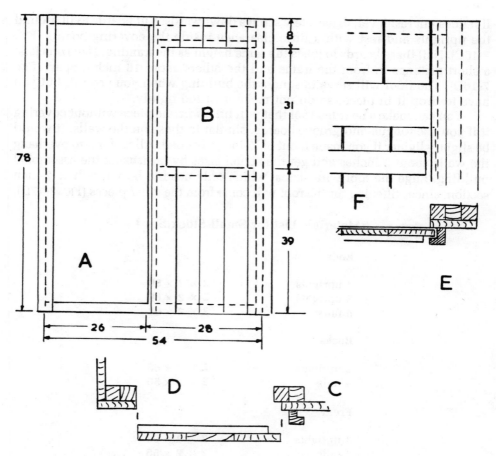

FIG. 7-18. *Details of the front of the small stock shed.*

the shutter at the desired height. Nail or staple wire mesh inside the window opening.

Make the door with tongue-and-groove or plain, vertical boards with three ledges across. With the ledges firmly attached, there should be no need for diagonal braces on this narrow door, but fit one if you think it necessary. Arrange the door to hinge on the outer edge and to overlap on the opposite upright, which will serve as a stop. At the top and bottom, allow for the door to overlap the frame, with plenty of clearance, particularly at the ground. Use two or three plain or T hinges and you can hold the other side with a turnbutton, unless you wish to fit a lock. Provide a wooden or metal handle.

Assemble the walls before making the floor or the roof. Coach bolts, ⅜-inch in diameter, at about 18-inch centers, should be satisfactory for the corners. Have the heads outside and nuts over washers inside.

Have the assembly square and on a flat surface when you make the floor, which rests on the bottom framing and should not need to be anchored down.

If it does not rest level, drive a few screws downwards. Cut boards to fit around the uprights and rest with a little clearance inside the covering boards (FIG. 7-16C). Nail these boards to joists the same height as the framing. Have one joist a short distance in from the frame and the others about 18 inches apart (FIG. 7-16D). The floor will serve to square the building when you erect it, but you have to drop it in place, so do not make it fit too tightly.

You can make a boarded roof that will lift off in one piece without covering it if you use tongue-and-groove boards, similar to those on the walls. It would be slightly lighter if you made it out of ½-inch plywood. Allow for it to overhang the walls about 5 inches and get the actual sizes by measuring the assembled building. Edge the roof with strips, which only need to be a 1-inch × 2-inch section, since stiffening for the roof will come from the other pieces (FIG. 7-17D).

Materials List for Small Stock Shed

Ends

2 uprights	2 × 2 × 80
2 uprights	2 × 2 × 68
8 rails	2 × 2 × 62

Backs

2 uprights	2 × 2 × 68
4 rails	2 × 2 × 56

Front

3 uprights	2 × 2 × 68
2 rails	2 × 2 × 56
4 rails	2 × 2 × 32

Shutter and door

3 ledges	1 × 6 × 26
2 ledges	1 × 4 × 26
2 guides	2 × 2 × 80

Floor and roof

5 joists	2 × 2 × 60
2 roof edges	1 × 2 × 72
2 roof edges	1 × 2 × 76
4 roof frames	2 × 2 × 60

Cladding

Tongue-and-groove boards	1 × 6

Fit 2-inch-square strips that will drop inside the walls. At back and front, there might be an easy clearance, but at the sides, make the strips fit fairly closely (FIG. 7-17E) as there will be bolts through to hold the roof down. Try the roof in position and drill holes for bolts into the side framing. The number of bolt holes depends on the situation—more are required if you experience high winds. In any case, there should be at least three, ⅜-inch coach bolts at each side.

At the back, make a flap that will hang down over the opening and swing up out of the way (FIG. 7-16E). You can make it with vertical tongue-and-groove boards, holding them with ledges across that will clear the opening when you close them. Put a strip across to take the hinges. To hold the flap up, put two wooden turnbuttons on blocks. Fit two more to hold the flap in the down position.

If you want a nest box, make it in the same way as you made the aviary hen house (FIG. 7-14C,D,E), but use the flap, just described, as the lid, then if the box is taken away, you can still seal the opening. If the nest box is to be temporary, take its ends far enough inside for two bolts to be put through at each place.

BARN

If you need to house larger animals, the building has to be bigger and stronger than those described so far in this chapter. A horse or other large animal or group of animals might put considerable strain on the structure, so it has to be substantial. You need a good barrier inside to spread any load on the walls. If the barrier is a strong, smooth lining, that reduces any risk of damage to the animals and makes cleaning easier. The building should be high enough to allow for good air circulation. These facts mean that if the building is to be adequate for its purpose, you have to be prepared to build fairly large.

The barn shown in FIG. 7-19 has a double-slope roof and double doors. It is 11 feet square and 10 feet high. Opening windows are high in the back and shallow windows are at the sides, above the lining. A building this size provides room for a horse to be stabled, with space for tack and feed. If you are concerned with smaller animals, you can accommodate two or more. The barn would also make a good place to store all the many things you would use on a small farm. The building has an attractive appearance, and you might wish to use it for many purposes in your yard. With different window arrangements, it would make a good workshop. You can alter doors to suit your needs. As shown, the doors are big enough for small trailers or other wheeled vehicles. A motor cycle, trail bike, or even a small car could fit through them.

The drawings and instructions are for a barn framed with 2-inch × 3-inch section wood covered with horizontal-shiplap boards (FIG. 7-20). Suggested lining material is particleboard or plywood. Measure your available space. Allow for laying a concrete base larger than the barn area. You must securely bolt down a building this size. You may prefabricate the ends and sides. Roofing is done in position after you have erected the walls.

Start by making the ends (FIG. 7-21). The back (FIG. 7-21A) is closed, but the front has a 7-feet-square doorway (FIG. 7-21B). Assemble framing with the

FIG. 7-19. *Make a strongly-built barn of traditional shape with double doors and windows at the side.*

3 inch way towards the cladding. Halve or tenon joints in the framing. Halve crossing parts of internal framing. The central rail is at the height intended for the lining. If that height does not suit your needs, alter its position. This height allows for shallow windows above the lining and under the eaves. Angles for the roof are shown (FIG. 7-21C). If you do not work exactly to these angles, it does not matter, as long as each end is symmetrical and they match.

Make the back and use it as a pattern for the outline of the front. Make the bottom rail of the front right across (FIG. 7-21D), but after you have erected the walls and anchored them down, you can cut it away to give a clear door opening. You can improve appearance of the ends if you add a broad filler piece at each corner (FIG. 7-21E). To allow for this filler piece, stop the shiplap boards over the center of the corner posts. Take them to the edges of the roof slopes.

The two sides are the same (FIG. 7-22A), unless you want to fit a side door or alter the number of windows. The drawing shows four window openings on each side (FIG. 7-22B), but you could reduce this to two or none. The tops of the windows are covered by the roof, so you cannot arrange them to swing

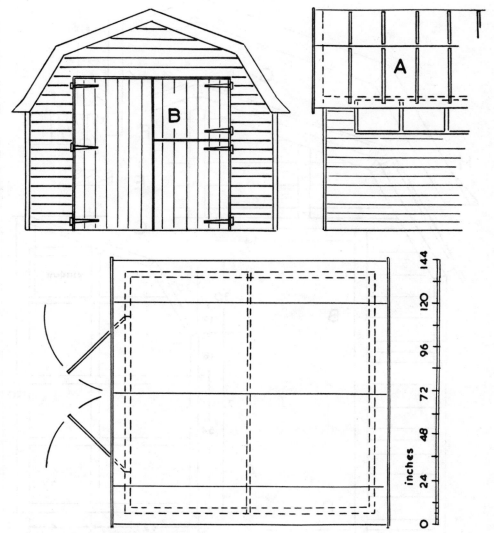

FIG. 7-20. Suggested sizes for the barn.

outwards very much. They could open inwards, but it probably will be satisfactory to make them fixed.

Allow for the cladding boards extending at the ends to cover the end-corner posts (FIG. 7-21F). Finish level at top and bottom. You do not have to bevel the top to match the slope of the roof.

You could add the lining at this stage or leave it until after erection. Take it to the window line (FIG. 7-22C). Cover it there with a sill extending outwards (FIG. 7-22D). Line the tops and sides of the openings (FIG. 7-22E). The roof will provide protection to the upper parts of the windows.

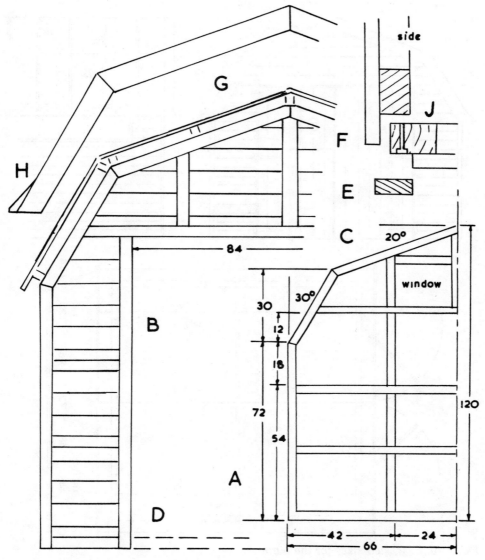

FIG. 7-21. *Details of the ends of the barn, which you should make first.*

The roof needs a truss halfway along. This truss must match the ends of the building, so use one of the ends as a pattern for the shape. The outline is the same down to the top of the side panels. As no boards are across to strengthen the framing, securely nail or screw gussets under the angles (FIG. 7-23A). The tie is the same height as the rail above the doorway. From its center, take struts at 45 degrees to it, up to the rafters (FIG. 7-23B). Cut the rafters to rest on the tops of the side frames, with locating blocks there (FIG. 7-23C). The purlins are 2 inch × 3 inch and the ridge is from 2-inch × 4-inch stock. Bevel the top of

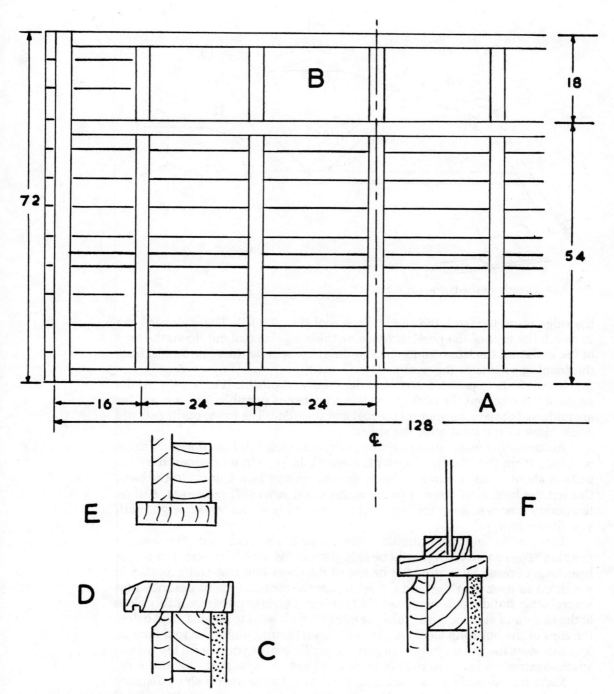

FIG. 7-22. A side of the barn and sections at the windows.

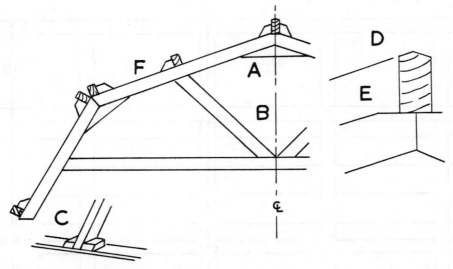

FIG. 7-23. Details of the barn-roof truss.

the ridge piece to match the slope of the roof (FIG. 7-23D). These slopes have to match the tops of the purlins. Measure their heights and cut down the tops of the ends and the truss, so the roof at the ridge will be the same height from the framing as it is at the purlins (FIG. 7-23E).

Make cleats to position and hold the ridge and purlins (FIG. 7-23F). At the angle of the roof, put the purlins as close together as possible. The top purlins are midway between the angle of the roof and the ridge. The lower purlins should come close to the joint with the sides.

Assemble the walls, using ½-inch bolts, sink their heads so the cover strips will hide them (FIG. 7-21J). For ample strength in any circumstances, have the bolts at about 12-inch centers. Check squareness and fasten down to the base. Cut out the bottom of the end frame under the doorway. If necessary, nail on temporary braces to keep the building square and hold the truss upright until you fit the roof.

The purlins and ridge should extend 6 inches at each end. Fix them in position. The roof covering could be shingles over ½-inch plywood, but 1-inch boarding, covered by roofing felt or any of the sheet-roof material supplied in a roll, taken over the ridge and turned under the roof boards, as described for several other buildings is suggested. Cut the board ends to meet reasonably close at the ridge and at the angles. Take the ends of the boards to about 1 inch below the tops of the building sides. A gap will be all around under the roof boards. You can leave the gap entirely or in part for ventilation, or you can fill the narrow spaces against the lower purlins or the wider gaps between purlins at the ends.

Make bargeboards at the end (FIG. 7-21G). Take the ends a short distance below the roof edges. You can give a traditional appearance with triangles if you turn the board line outwards (FIG. 7-21H). Put battens down the slope of the roof over the covering at about 18 inch intervals (FIG. 7-20A).

The double doors are ledged and braced, but because of their size and weight, you must double them around the edges (FIG. 7-24A). Put lining strips around the door opening, covering the wall lining as well as the cladding (FIG. 7-24B). Make the doors with vertical tongue-and-groove boards. Put braces across, level with the board ends, and fill in to the same thickness at the edges (FIG. 7-24C). So the diagonal braces take any compression loads which come on them without allowing movement, fit them closely at their ends, making sure there are no gaps which might cause the door to drop.

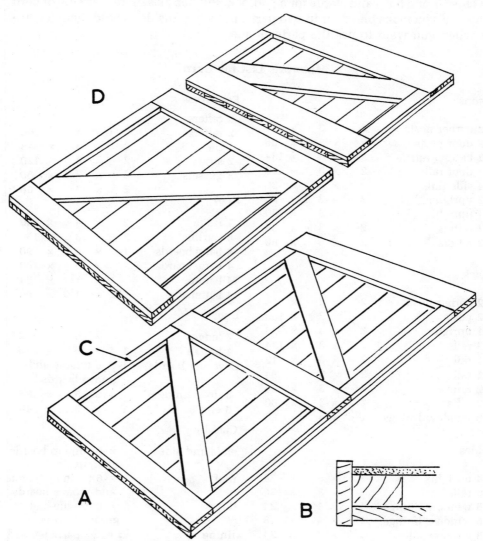

FIG. 7-24. Construction of the barn doors.

You could make one half in two parts for the usual stable door pattern (FIG. 7-20B and 7-24D). A height of 48 inches would give a horse about 36 inches to put its head through, but the gap would not be big enough for most animals to jump through. Make each door part similar to the large door, with bracing upwards from the hinge side.

T hinges about 18 inches long would be suitable for hanging the doors. Notch the lining strips around the doorway for the hinges, which should come over the ledges on the doors, and be held with long screws in both parts. You can take bolts right through to nuts instead of screws. Arrange bolts upwards and downwards on the inner edge of one door and place a lock on the other door to close it or a hasp and staple for a padlock. Put handles on the outside of both doors. If you make one door in two parts, put a bolt inside to hold them together when you want to use the parts as one.

Materials List for Barn

Front

2 corner posts	2	× 3	×	74
2 door posts	2	× 3	×	86
1 bottom rail	2	× 3	×	134
1 door rail	2	× 3	×	120
4 side rails	2	× 3	×	26
2 uprights	2	× 3	×	26
1 upright	2	× 3	×	36
2 rafters	2	× 3	×	84
2 rafters	2	× 3	×	60

Back

2 corner posts	2	× 3	×	74
2 posts	2	× 3	×	100
1 upright	2	× 3	×	36
3 rails	2	× 3	×	134
1 rail	2	× 3	×	12
1 rail	2	× 3	×	54
2 rafters	2	× 3	×	84
2 rafters	2	× 3	×	60
8 window linings	1	× 4	×	26

Sides

14 uprights	2	× 3	×	74
8 rails	2	× 3	×	130
8 window linings	1	× 4	×	24
16 window linings	1	× 4	×	18
8 window sills	1¼	× 5	×	24
4 corner fillers	1	× 2½	×	74

Roof truss

2 rafters				
2 rafters	2	× 3	×	60
1 tie	2	× 3	×	48
2 struts	2	× 3	×	120
3 gussets	2	× 3	×	60
	2	× 3	×	36

Roof

1 ridge	2	× 4	×	150
8 purlins	2	× 3	×	150
4 bargeboards	1	× 6	×	90
4 bargeboards	1	× 6	×	80
14 battens	½	× 1½	×	84
14 battens	½	× 1½	×	60

Doors

7 ledges	1	× 6	×	42
4 braces	1	× 6	×	70
Covering boards	1- × -6 tongue-and-groove boards			
3 door linings	1	× 5	×	86
8 edges	1	× 3	×	40

Cladding

ends and sides	1- × -6 shiplap boards or equivalent
roof	1- × -6 plain or tongue-and-groove boards
doors	1- × -6 tongue-and-groove boards
lining	½ or ¾ particleboard or plywood

How you make the windows depends on the use of the barn. If you want it to be weathertight, the windows should be made closely. The overhang of the roof, however, gives partial protection to the windows and you can use a simple construction if a slight risk of leakage is not important. With the window openings lined, you can hold glass in between double strips (FIG. 7-22F). You could embed the glass in jointing compound. Another way would be to put single strips around the glass and putty the glass against them.

For better windows, frame them separately to fit in the openings (FIG. 5-4). If you want any windows to open by being hinged at the top, make them this way.

STABLE

You could use the barn just described for a horse, but it is a general-purpose building with several other uses. If you want to build a stable, you can include features to suit that purpose only. This stable is designed to suit one fairly large horse, with an adjoining storage space for feed, tack, and all the equipment associated with keeping a horse. For more than one horse, you can extend the building two or three times. For a pony, you can reduce the sizes slightly.

The stable shown in FIG. 7-25 and 7-26A has accommodations which are 11 feet × 12 feet, and there is a separate section built in alongside which is 6 feet × 11 feet. Both parts are under the same roof, which extends to make a 4-foot-wide canopy at the front, so the area covered by the building is about 17 feet square. A concrete base should extend several feet beyond this and you must make provisions for anchoring the building securely. Besides any effect of wind, a horse which throws its weight against a wall, could move an insecure

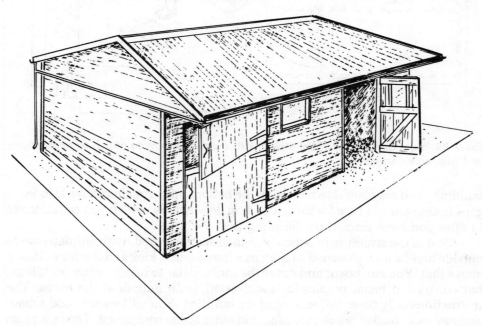

FIG. 7-25. This stable has a store or tack room built in alongside.

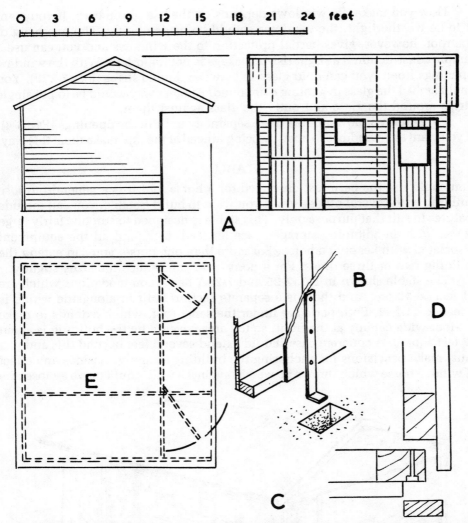

FIG. 7-26. Sizes and layout of the stable with details on how to attach to foundations and assemble corners.

building. You may bolt down through the bottom rails. An alternative is to leave gaps in the concrete base for bars extending downwards, which you can concrete in after you have erected the building (FIG. 7-26B).

Most of the structure is 2-inch × 3-inch section wood, with shiplap boards outside and ¾-inch plywood as high as a horse might kick and a thinner lining above that. You can board and cover the roof similar to many earlier buildings, but corrugated metal or plastic is suggested, with gutters at the eaves. The instructions only cover the making of the building. You will want to add a manger, hay rack, bucket holder, tie ring, and other stable equipment. The tack room will require various racks and a table, much of which you can build in. You can leave the roof area unlined with gaps at the edges for ventilation, or you could seal it only, or you could attach lining sheets as well under the purlins. You can use the roof area for storage, however, you could fit a flat ceiling in the tack room for comfort.

Although the building is large and some sections are heavy enough to require help in making and erecting them, construction is simple and very similar to some of the smaller buildings. If you can make sections flat on the concrete base, they should be easy to move into place for erection.

The building is shown without windows except at the front, which is what many horse owners prefer. You also could have windows high in the back or end of the stables and above bench level in the tack room, fitting them into spaces in the framing and making them as described for the barn and other buildings. The following instructions allow for a window beside the stable door and one in the door of the tack room.

The pair of ends and the division between the stable portion and the tack room are almost the same. Make one end and use it as a guide when making the other parts. Also use it to check the height of the back. All framing is 2-inch × 3-inch wood with the 2-inch side towards the cladding.

Make the framing with halved joints where parts cross, but at the outside, you could notch or use tenons. Space upright evenly in the width (FIG. 7-27A) and in the height to the eaves. This spacing should suit cladding with shiplap boards or vertical tongue-and-groove boards. At the eaves, drop the main horizontal member 4 inches (FIG. 7-27B). Mortise-and-tenon joints are advisable here, even if you use other joints elsewhere. This arrangement allows simpler and stronger eaves joints than if the parts met at the same level and it provides for a board at the edge of the canopy.

Although firmly-fixed cladding will help to brace the assembly, it might be advisable to include sway bracing (FIG. 7-27C) in two panels.

Fit the cladding on the ends to the edges of the roof portion. At the corners stop it at half the thickness of the uprights, to allow for filler strips (FIG. 7-26C).

You can line the ends now, allowing for where the other uprights have to fit, or leave it until after you have erected the building. Use ½-inch or ¾-inch plywood or particleboard from the floor level to the rail, 68 inches up. If you intend to line with lighter material above this point, stop the thick lining at half the thickness of the rail to allow space for nailing the additional lining above it. Stop the lining on the edges of the corner uprights.

The framing of the division is the same as an end, except you must cut the rear edge back by the thickness of the back framing (3 inches) to fit inside (FIG. 7-27E and 7-28A) and the front upright must fit similarly (FIG. 7-27F). Line the framing on both sides. Although you could add some of the lining at this stage, you should fit it into the back and front for the best finish (FIG. 7-28B). It is easier to fit the lining after erection when you have the parts bolted together.

The back (FIG. 7-27D) settles lengthwise arrangements. If you want to alter them, do it now. Fit one upright against the division. Space other uprights evenly. Include some sway bracing. Height should match the ends, but you do not need to bevel the top edge to match the slope of the roof.

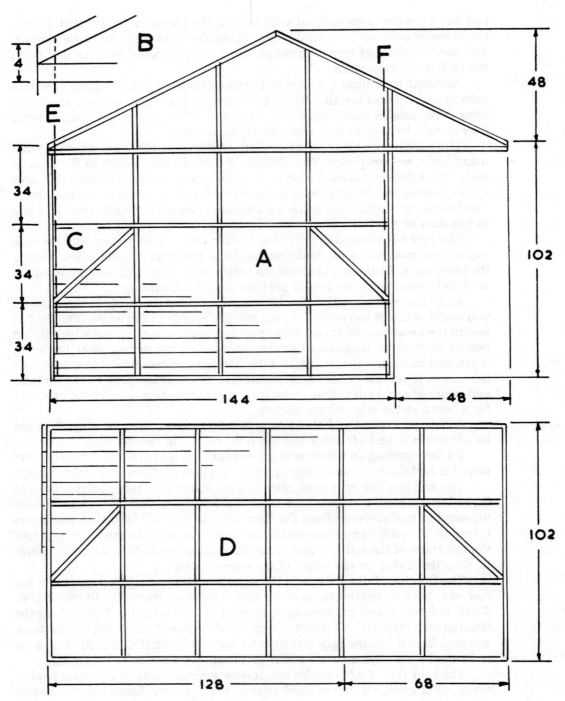

FIG. 7-27. An end and back of the stable.

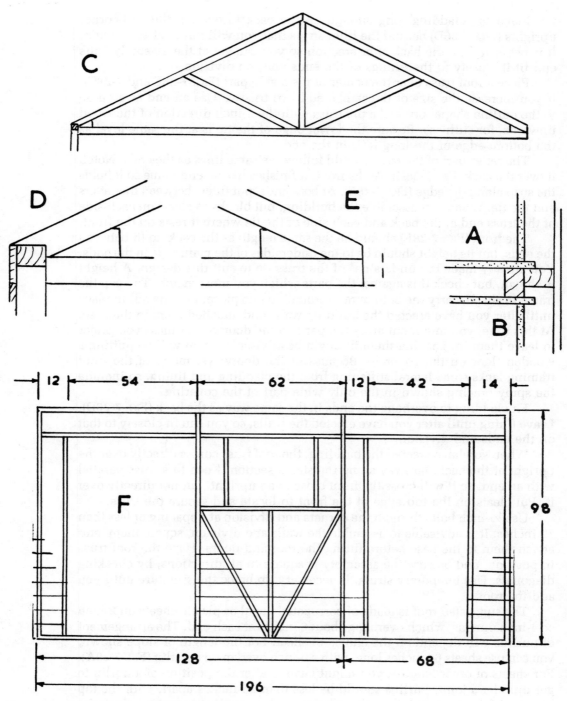

FIG. 7-28. A roof truss and front of the stable.

Make the cladding long enough on the back to overlap the end-corner uprights (FIG. 7-26D) behind the filler strips that you will put in after erection. It is better to line the back after erection so you can get at the assembly bolts and fit it closely to the linings of the ends and the division.

Place a roof truss over the center of the stable part (FIG. 7-26E and 7-28C). If you increase the size of the stable, add two trusses. Use an end as a guide to the outline shape, but make the truss with the 3-inch direction of the wood upwards, for stiffness. Be sure the lower edge of the tie is at the same level as the bottom edge of the long rail, in the end.

The rafter part of the truss should follow the same lines as the ends. Notch it over the back (FIG. 7-28D). At the front, it finishes like an end frame so it holds the strip along the edge (FIG. 7-28E). At both ends, put fillers between the rafters and the tie. When you assemble the building, put blocks as cleats on each side of the truss end at the back and each side of the tie where it rests on the front.

The front (FIG. 7-28F) should be the same length as the back, to fit between the ends, but its height should be to the undersides of the main rails in the ends. You already made the undersides of the truss tie to suit this design. A height is shown, but check this against the parts which you must match. To keep the frame in shape, carry the bottom rail through in one piece. Let this rail in place until after you have erected the building walls and attached them to the base. At this time, you might cut away the parts in the doorways, unless you prefer to leave them in. Leaving them in might be advisable if you will be putting a wooden floor on the concrete. Because of the doorways, much of the front framing lacks some lateral stiffening from the cladding and lining, so include the spray bracing shown in the only wide part of the covering.

Fit cladding to overhang the ends in the same way as the back (FIG. 7-26D). Leave lining until after you have erected the walls, so you can fit closely to that on the ends and division.

When you have erected the building, the roof truss comes directly over the upright at the back, halfway along the stable section. Keep this truss parallel with an end, so it will cross the front close to an upright, but not directly over it. Put cleats on the top edge of the front to locate and secure the truss.

Use ½-inch bolts through the corners and division at a spacing of less than 19 inches. It is advisable to assemble the walls and division, square them, and attach them to the base before fitting the roof and doors. Have the roof truss in position, and be sure the assembly is square in all directions, by checking diagonals. Use temporary struts, if necessary, to brace the structure until you add the roof.

The suggested roof is made of corrugated metal or plastic sheets on 2-inch × 4-inch purlins, which overhang about 6 inches at each end. The arrangement of purlins will depend on the choice of sheet. For the length of slope shown, you can use sheets 60 inches long, with a 6-inch overlap on a purlin (FIG. 7-29A). For sheets of other lengths, you might have to alter the position of a purlin to get it under a joint. Purlins should be less than 25 inches apart, with the top ones as high as possible and the other one close to the eaves. Use cleats for location and fixing (FIG. 7-29B). Attach the sheets with large-headed nails or

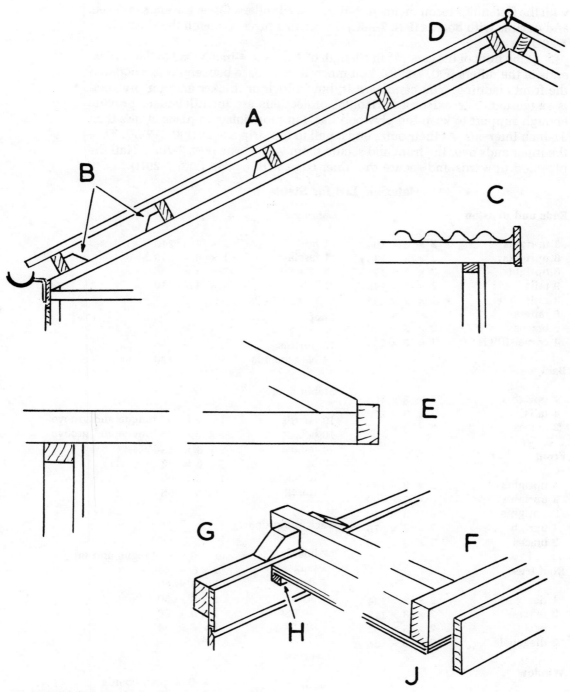

FIG. 7-29. *Roofing details of the stable.*

with the fastenings recommended by the sheet suppliers. Cover the ends of sheets and purlins with boards (FIG. 7-29C). Fit a ridge piece to match the sheets (FIG. 7-29D).

At the front of the canopy, fit a length of 2-inch × 4-inch wood to the vertical ends of the rafters (FIG. 7-29E). You must fill the space between this wood and the front cladding. You can board it, but ½-inch or thicker exterior plywood is suggested. The existing horizontal projections are insufficient to provide enough support to keep the plywood flat. Put more joists in place at less than 31-inch intervals. At their outer ends, nail to the strip across (FIG. 7-29F). Take the inner ends over the front and secure them with cleats (FIG. 7-29G). Nail the plywood upwards and secure the inner edge with a strip (FIG. 7-29H).

Materials List for Stable

Ends and division

3 uprights	2 × 3 × 104
6 uprights	2 × 3 × 130
6 uprights	2 × 3 × 145
9 rails	2 × 3 × 146
3 rails	2 × 3 × 194
6 rafters	2 × 3 × 112
6 braces	2 × 3 × 50
4 corner fillers	1 × 3 × 104

Back

7 uprights	2 × 3 × 104
4 rails	2 × 3 × 200
2 braces	2 × 3 × 55

Front

4 uprights	2 × 3 × 100
3 uprights	2 × 3 × 90
2 uprights	2 × 3 × 30
1 upright	2 × 3 × 70
2 braces	2 × 3 × 80

Roof truss

1 tie	2 × 3 × 194
2 rafters	2 × 3 × 112
1 post	2 × 3 × 52
2 diagonals	2 × 3 × 56

Window

4 linings	1 × 6 × 30
8 fillets	1 × 1 × 30

Canopy

1 rail	2 × 4 × 210
1 fascia	1 × 6 × 210
5 joists	2 × 3 × 56
1 fillet	1 × 1 × 210

Roof

10 purlins	2 × 4 × 220
4 bargeboards	1 × 6 × 120

Stable door

10 boards	1 × 6 × 60 tongue and groove
10 boards	1 × 6 × 45 tongue and groove
4 ledges	1 × 6 × 56
2 braces	1 × 6 × 72
2 linings	1 × 6 × 90
1 lining	1 × 6 × 56

Tack room door

7 boards	1 × 6 × 90 tongue and groove
3 ledges	1 × 6 × 44
2 strips	1 × 3 × 40
1 brace	1 × 6 × 50
2 linings	1 × 6 × 90
1 lining	1 × 6 × 44

Covering

cladding	1- × -6 shiplap boards
lining	½ or ¾ plywood or particleboard
canopy	½ plywood

Put a fascia board along the front (FIG. 7-29J) to cover the edge of the plywood. Fit a gutter with brackets to this board. This gutter might be led across an end to join the rear gutter at one downpipe, or it could have its own downpipe to a soakaway or a water barrel.

Fit lining pieces around the doorways and the windows, to cover the cladding and lining (FIG. 7-30A). The canopy protects the window, so there would be no need for a wider sill.

Make the two-part stable door in the same way as described for the barn door (FIG. 7-24D), but line both parts in the same way as the walls. Hang it with similar T hinges and fit a strong fastener or lock or hasp and staple for a padlock. Have a sliding bolt between the two parts. Fit a catch to hold the upper part, or the two combined, in the open position.

Some horses might chew the top edge of the lower door. Wrap and nail a piece of sheet metal over the edge (FIG. 7-30B) to protect it.

The tack-room door does not have to be quite as substantial, but it should be strong—if it is light enough to flex, you might crack the window. Make this door with three ledges. Space two to suit the size window you want and arrange

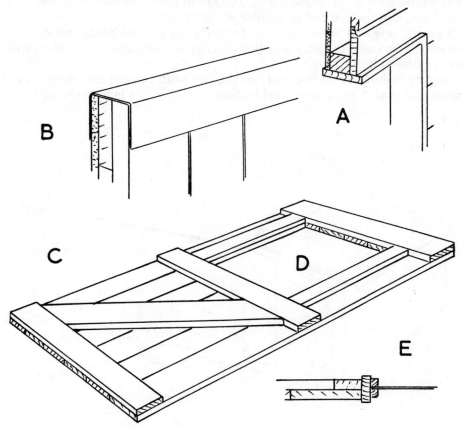

FIG. 7-30. Door and doorway details for the stable.

a closely-fitting brace in the panel below (FIG. 7-30C). Put a lining strip around the window opening and mount the glass between two fillets (FIG. 7-30E). Nail them in so they are removed easily if you have to replace the glass. Hinges could match those on the stable door and you probably will want to fit a lock.

You can fit the window in the stable in the same way or you could arrange an opening window similar to those of the barn.

DOG KENNEL

The common dog kennel with an end doorway does not provide much protection when the wind blows rain in the direction of the door. If you make it as a one-piece building, cleaning is difficult. The kennel shown in FIG. 7-31 has a porch, a raised floor, and a roof which lifts when you want to get at the inside to clean or for other reasons. The sizes suggested in FIG. 7-32 should suit a dog of medium size, but you can modify them easily if you think your dog would be happier with a different size.

Construction is with ½-inch-exterior-grade plywood on a frame of 1-inch × 2-inch strips. Covering could be with tongue-and-groove boards, or shiplap boards would look good on the walls. You could paint the roof, or it might be better if you use roofing felt or similar material on it.

It probably would be satisfactory if you merely nail all joints, but you can halve some. Use water-resistant glue between the plywood and the framing, with plenty of fine nails.

Start by making the closed end and the part with the entrance hole. At the entrance end (FIG. 7-33A), notch the bottom corners so the lengthwise rails can

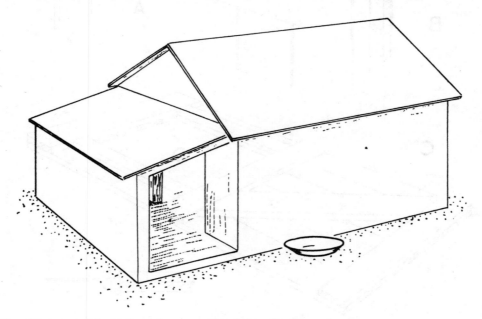

FIG. 7-31. A porch shelters the entrance of this dog kennel.

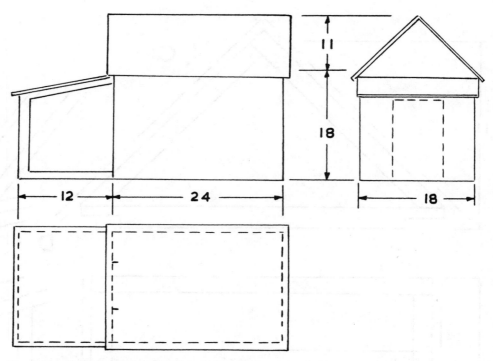

FIG. 7-32. Suggested sizes for the dog kennel.

go through to the porch (FIG. 7-33B). You could do the same at the top corners, although it probably will be sufficient to rely on the plywood holding the lengthwise rail in place. Cut the plywood at the bottom notches. At the doorway, round all edges of the frame and plywood. Make the closed end similar to the end with the entrance, but without the door framing. Do not notch the plywood at the corners.

The gable ends (FIG. 7-33C) have a 45-degree slope and are as wide as the other parts over the plywood sides. Notch the framing and not the plywood for the lengthwise parts (FIG. 7-33D). Check that the width matches the other parts and the two gable ends are a pair.

Make the two, bottom side rails (FIG. 7-34A) with halving joints to match the ends. For the open side, join on plywood, extending enough to overlap the plywood on the ends, with a strip to fit between them at the top. You can halve this strip if you wish (FIG. 7-34B). Make a longer piece of plywood for the closed side (FIG. 7-34C), cut down by the thickness of its roof. You can do the framing of the edges of this part when you assemble the kennel by adding the porch end.

The porch end (FIG. 7-34D) is the same width as the main ends and is as high as the sloping side. Notch the bottom framing to match the lengthwise strips. Place the top-crosswise piece edgewise, and bevel the top part to suit the slope of the porch roof.

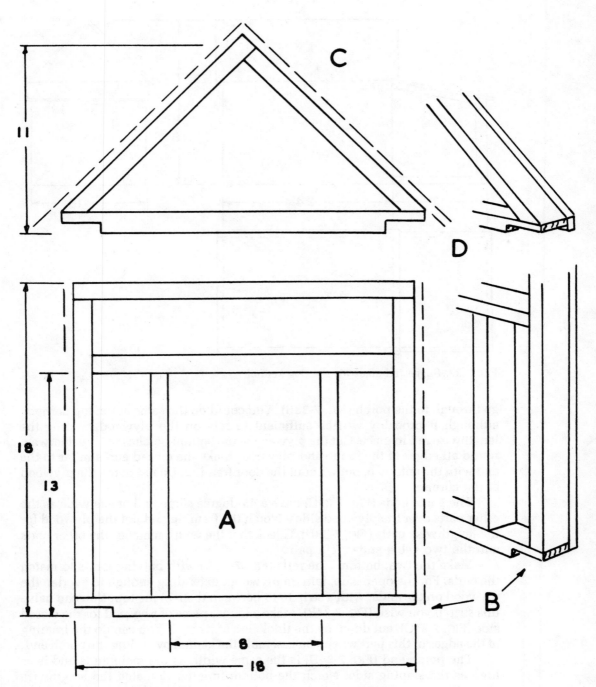

FIG. 7-33. End parts for the dog kennel.

Assemble all the parts made so far. Add a stiffening piece to the top edge of the porch back and the porch front. Make a bottom of plywood to rest on the bottom frame strips. To get the bottom in, you will have to make a joint across the doorway. You can leave it loose, so you can remove it for cleaning, but if you want to make it insect-proof, embed it in glue and nail it down. The gluing and nailing also will contribute to strength and keep the kennel in shape if the ground underneath is uneven.

Nail a strip above the entrance hole to support the inner end of the porch roof, with its top sloping to suit. Put on the plywood porch roof, placing it close to the main end and overhanging the walls by 1 inch. Notch it under the main roof. Round the exposed edges and corners, particularly if you will not be covering this roof.

Make the two gable ends into a roof by joining their bases with lengthwise strips so they match the tops of the walls. At the apex, put in another lengthwise strip (FIG. 7-34E) to support the roof covering. The usual ½-inch plywood should be stiff enough to hold its shape on a roof of this size, without intermediate support, but if you think it needs stiffening, include some central rafters.

Cover the roof with plywood to overlap 2 inches at the ends and eaves. Round the edges and corners if you are not covering the roof. If you cover it with felt or other material, turn the edges under and tack them. Put light battens down the slopes, near the ends, and one or two intermediately.

Materials List for Dog Kennel

Ends

9 strips	1 × 2 × 20
2 strips	1 × 2 × 16

Gables

2 strips	1 × 2 × 20
4 rafters	1 × 2 × 16

Porch end

2 strips	1 × 2 × 20
2 strips	1 × 2 × 15

Lengthwise parts

2 strips	1 × 2 × 38
2 strips	1 × 2 × 24
2 strips	1 × 2 × 14
3 roof strips	1 × 2 × 26

Covering

	½ plywood

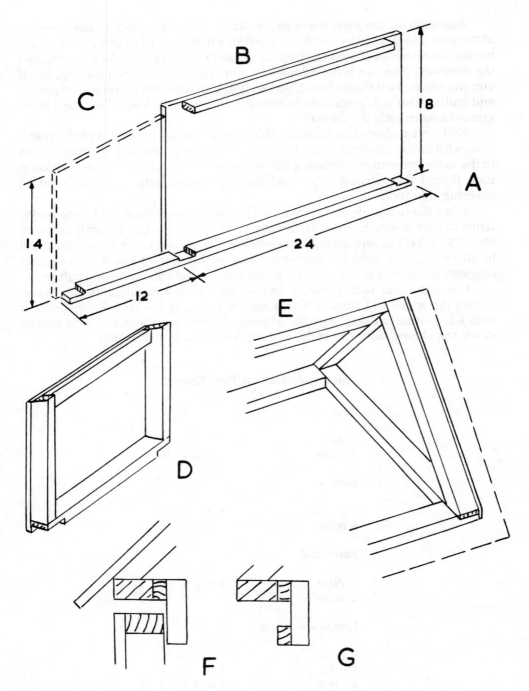

FIG. 7-34. Sides, ends, and roof of the dog kennel.

Check the fit of the roof on the walls. It has to be given a positive location, and you might wish to fasten it down, otherwise it might blow off. At each inside corner of the roof, screw 1-inch × 2-inch strips projecting downward and packed out so they fit easily inside the top of the walls (FIG. 7-34F). At one side, add ends to hook under the top edge of the top strip of the wall (FIG. 7-34G). At the other side, on the outside, put screw hooks and eyes near each end to hold the roof down. You can use some lever-action clips and other fasteners there as alternatives.

POLE BARN

If you have poles of various sizes available, it is possible to make several type buildings with them. You make these buildings similar to the ones used by early settlers who had to adapt their needs to the available material—felled trees which they used in the round if possible.

A pole barn can have a useful capacity, and it will blend into the scenery better than most other structures. Even if you are not a farmer, the building can have many uses on your property as well as being an example of traditional assembly which should be pleasing to the eye. The pole barn shown in FIG. 7-35 is of modest size, although you can use the technique for a barn of very different dimensions. In its width, you can divide it into three bays of about equal size. You can let the center bay open right through, but divide the side bays into stalls

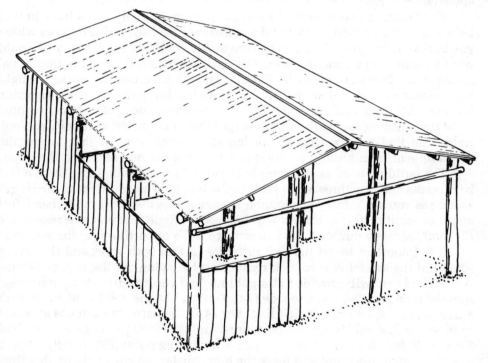

FIG. 7-35. This pole barn has many parts made from natural wood.

for cattle or horses, or use them for smaller animals. You can build walls which are full height or you can stop them at a lower level. Plenty of space is available for storage of feed, fertilizer, and other things, including some implements.

If you want to produce a barn of traditional appearance, walls and partitions should be of split poles, but you can back these with squared, wooden rails or even plywood, if the walls must be close-fitting and free from draughts. Although you can cover the roof in several ways, you have a choice of boarding or using plywood which you would cover with shingles or you could use corrugated plastic or metal sheets. In this example, it is assumed that you will use corrugated sheets, but you can adapt the design to other forms of roofing.

The choice of poles will depend on what is available, but for uprights, the poles should be straight and about 6 inches in diameter, no matter what wood you choose to use. Lengthwise shakes are almost inevitable in many softwoods and it does not matter if they are small. For rafters, you can use straight poles, if you have them. You can flatten poles slightly out of true on their upper surfaces, providing it does not weaken them too much. They should also have an average diameter of 6 inches. If suitable poles for rafters are not available, you might have to use sawn wood in 2-inch × 6-inch sections. You might be able to use poles at the ends of the barn where they show and sawn wood intermediately. With modern facilities, it obviously makes sense to use modern materials and techniques, even if your aim is for a traditional external appearance.

The building could have some sway bracing, and any walls built in will help to maintain rigidity. Most of the stiffness of the barn is due to posts which you buried in the ground. This type structure is not really a building you would want to mount on a concrete base. The concrete would destroy the traditional appearance. The site should be flat and should have compacted soil. Slight unevenness will not matter, but you might wish to bring the area to a reasonably level condition. If you decide to accept more unevenness, do not be tempted to let the building follow the slope—a roof that is not horizontal does not look right. See that the roofline in the length is level, even if that means the measurement from ground to eaves is 2 feet more at one end than the other.

The building used as an example in FIG. 7-36A is divided into three 8-feet bays across and into three 8-feet parts in the length. You could make the length as long as you wish, but you should not make across members more than 8 feet apart, or additional roof trusses will be needed. Eight feet makes a reasonable division for many purposes, so you arrange walls or partitions at these points.

It is important to get the upright poles in line both ways and the layout square, if the assembly is to look right. Start by laying out the post positions. A stretched cord will provide a straight line, and you can drive temporary pegs into the pole positions on one side (FIG. 7-37A). Use the 3:4:5 method to mark a line of pegs square to this first line. For 24 feet square, you can use an 8-feet unit, so the first side is 3 units. Measure 4 units (32 feet) from one end peg and 5 units (40 feet) from the other end peg to a meeting point (FIG. 7-37B). Stretch a cord through this point and mark the 8-feet distances with pegs on this line.

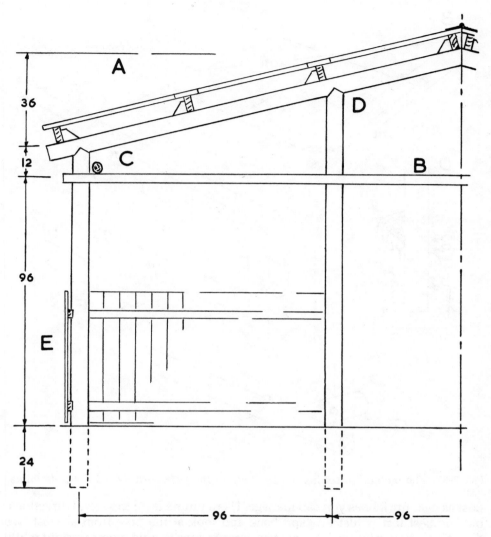

FIG. 7-36. An end view of half the pole barn.

For the lines of other pegs, measure parallel to these first square lines (FIG. 7-37C).

How deep you sink the posts depends on the soil, but for rigidity, they should go in at least 24 inches, even in the most dense soil. It might be possible to drive stones tightly under and around a post, but it will be better to embed a post in concrete, which will spread the load (FIG. 7-37D). The concrete may come to the surface, or it could just form a foot, with stones and soil compacted above.

Have the posts longer than they will have to be, so you can trim the tops to match when they are all in position. At this stage, it is important to get each

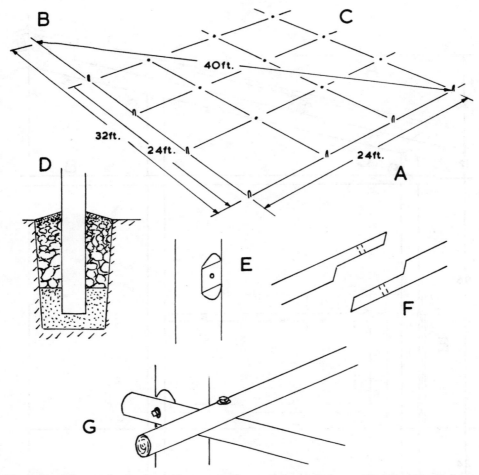

B

C

40 ft.

32 ft.

24 ft.

24 ft.

A

D

E

F

G

FIG. 7-37. *The method of setting out the base of the pole barn. Details of pole joints.*

post as near upright as you can manage. Use a plumb level in several directions, but the best test is just to stand back and look at the pole from at least two directions. You might need some temporary props to hold a post upright while the concrete sets and until you attach other parts to it to keep it in shape.

When you have one corner post in position, it probably will be wisest to work the corner diagonally opposite it next, then the other corners. As you progress, you can sight one post against another. As you get more posts up to your satisfaction, it is easier to sight the remaining posts and get them true.

It will help to stabilize the assembly if you fit the horizontal ties next (FIG. 7-36B). They only need to be about 3 inches in diameter. They cross 8 feet from the ground, but if the ground is uneven, treat this as a minimum and keep the ties level. Use a cord across with the aid of a large spirit or other level and mark the crossing points of one set of uprights. At each place, make a flat surface, but do not go very deep (FIG. 7-37E). Make matching flats on the tie. If you do not have a pole long enough to go right across, you can make a joint over a post (FIG. 7-37F). Use a bolt which is at least ⅝ inch in diameter at each crossing, with large washers under the head and nut.

To hold the assembly in shape lengthwise, put more ties on top of the first set close to the eave's joints (FIG. 7-36C and 7-37G). Again, arrange meeting flats and bolt through. You will maintain stiffness on the inside uprights if you join partitions and other parts to them. If you think it advisable, fit more lengthwise ties near them.

It is important that the roof is straight, level, and without twist. If some of the lower parts of the building are slightly out of true, it might not matter and might add to the picturesque appearance of the structure. Errors in the roof will be more obvious and will detract from appearance, even if it is weatherproof.

Before cutting off any uprights, establish a horizontal line all around the building. The positions of the main ties will serve as a guide. The line can be near their level and about 12 inches down from where the eaves will be. Stretch a cord along the uprights of one side, and with the aid of your level, make it horizontal. Mark this line prominently on each post. Go around a corner and do the same across an end, starting at exactly the level of the first line. Place a similar line across the other end, then when you draw a line on the other side, these lines at the corners should be level. You might not achieve perfection, but if you are much more than 1 inch out, go back and check what you have done.

From the datum marks, measure up to where you will cut the tops of the outer posts. At one end, use two straight pieces of wood long enough to overlap at the apex of the roof. Put them in the rafters position, with their lower edges where you intend the lower edges of the pole rafters to be. Clamp their overlap and either clamp or temporarily nail to the posts. When you are satisfied with the shape, mark on the poles where the guidelines are to be, then remove these pieces.

You can use several methods to join the rafters to the posts. The simplest way is to put the rafters alongside the posts, with flat sides against the posts and bolt through them, similar to the joints of the ties. This method allows experiments with levels, before you drill through for bolts. The method shown has notched joints. Each post has a tapered top fitting into a notch in the rafter. Arrange this fitting to go less than halfway through the rafter (FIG. 7-36D and 7-38A). At each joint, drill downwards and drive in a ¾-inch-diameter steel rod as a dowel (FIG. 7-38B). At the apex, halve the rafters together, with two ¾-inch bolts (FIG. 7-38C). At the eaves, cut off the rafters 9 inches outside the posts.

Assemble the rafters at one end, then use them as a guide when you erect the others. Sight along the top surfaces and check that they are as in line as is possible to get with natural poles. If there is a pronounced curve or kink in any rafter pole, you may cut it level, but do not take away too much wood. It might be better and easier to only cut away at purlin positions.

At the sizes suggested, it is possible to cover each slope with three rows of 60-inch, corrugated-metal sheets overlapping 9 inches. Locate the purlins to suit the sheet joints. With different length sheets, alter the purlin positions to suit. If you are going to board the roof, purlins may be 48 inches to 60 inches apart. On each rafter, arrange the top purlins quite close to the ridge. Position the lowest one within a few inches of the end of the rafter. The position of others is best found by experimenting with a row of corrugated sheets, so the overlap will come where you can nail centrally into the purlin.

The purlins could go right through, if you have wood long enough, but they will be easier to handle if you overlap them on the intermediate rafters (FIG. 7-38D). At the ends, they can project by whatever amount you want the roof to overhang—18 inches is reasonable. Where the purlins cross rafters, fit cleats. With 2-inch × 6-inch purlins, each cleat should be at least 4 inches high (FIG. 7-38E). Flatten the rafter to make a good bearing for the cleat and purlins (FIG. 7-38F). If there are no problems with levels, this flattening can be quite slight, but if you need to make any adjustment to the purlin level, you need to cut deeper or put a packing between rafter and purlins.

Sway bracing will aid rigidity, particularly if the barn is in a windy position. In two bays on each end, nail diagonal pieces upwards into two rows of purlins. These could be 2-inch × 3-inch sections. If you build walls into the barn, diagonals between rows of rails also will help resist wind loads.

Cover the roof and put a top on the ridge. At the eaves, extend the sheets a short distance below the purlin. You can then fit a gutter with brackets to the purlins (FIG. 7-38G). You could fit bargeboards to the ends, but you can finish this type of building without them.

How you arrange the accommodation of the building depends on your needs. In its simplest form, you can leave the building as it is, and use it as a shelter for implements and other equipment. At the other extreme, you may completely encase it and fit doors. You can put walls to the full height or only part way up the sides.

The arrangement of posts inside will allow the easy erection of partitions at 8-feet intervals, which should be right for horse or cow stalls, or storage compartments for grain, feed, and other things in containers. You do not have to treat the whole building in the same way. You could have walls to the eaves on one side and no walls, or just low ones, at the other side. Similarly, divisions could go to the roof or only part way up the inner posts. A useful arrangement allows the middle bays to be clear right through, with accommodation of various sorts to suit your needs along each side.

You can cover walls with plain boards, arranging them vertically or horizontally. If you want to maintain a traditional appearance, external walls should be split poles or the slabs which come off a log when it is squared in a saw mill. These slabs have the bark on. It is advisable to peel the bark off, as it harbors insects and might encourage rot. The natural outside of the wood still will have a suitable appearance. If you want the closest fit between slabs, cut the edges parallel (FIG. 7-39A). Rails to support the covering pieces may be 2-inch × 4-inch-sawn pieces, notched into the posts (FIG. 7-36E and 39-B). If you are making a wall extending over several bays, adjust the depths of notches so the rails are straight and parallel.

Make partitions the same way, but you can use plain boards. If animals are involved, the boards should be at least 1 inch thick. You might want to put boards

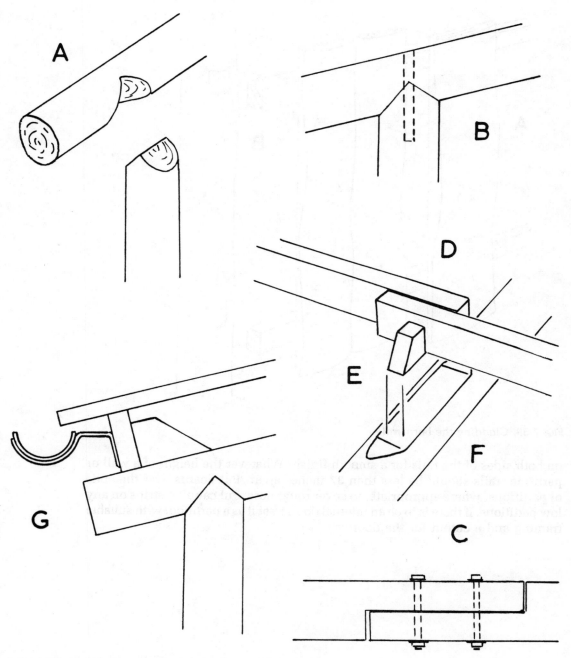

FIG. 7-38. Methods of joining parts of the pole barn.

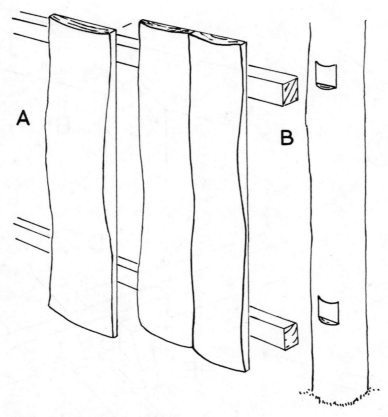

FIG. 7-39. *Cladding the barn walls.*

on both sides of the rails for a smooth finish. Whatever the height of a wall or partition, rails should be less than 37 inches apart. Put boards over the ends of partitions, where appropriate, to cover roughness. Put capping strips on any low partitions. If there is to be an internal door, treat it as a partition, with suitable framing and a cutout for the doorway.

8

CHILDREN'S BUILDINGS

CHILDREN AT PLAY LIKE TO HAVE SOMETHING TO GET INTO. THEY MIGHT IMPROVISE shelters or use their imagination with the most improbable things. If you can provide them with an actual building to play in, they might go into raptures and will be occupied for a long time. They can use such a house of their own for many hours of make-believe. They can furnish and decorate it. They will use it for many occasions over a long period. They do grow up, however, and no longer have a use for the building. An empty, unused building can be a problem, unless other children come along to take over. For the usual small family, any sort of play building should not be permanent.

You must consider if the building is to be permanent, or if it should be made portable, so it does not have to stay outside when you no longer need it. If it is to be portable, where will you store it? If it is portable and folds, you might be able to place it against a garage wall or other place that will take length and width, and a reduced thickness.

A play building or "Wendy" house usually will be much smaller than a building used by adults. This feature could be one of its attractions—children can get into it, but their parents cannot. This design means that you can use lighter construction and the work is easier to handle, even if your shop is small and your tools and equipment are not very plentiful. You might want to use the building for only a few years life, so it does not have to be very durable. You could use framed hardboard for a very economical building, but it might damage easily and it would deteriorate rapidly under moist conditions. Exterior-grade plywood is very suitable, as panels usually are smaller than a standard sheet. For many parts, plywood might be stiff enough without added framing.

If the building is to be permanent and must withstand any weather, you will have to make it more like a larger building, as described earlier. You might consider making it a combination building, even if it is small. Design the front to appeal to children, but place a large door at the back which will allow you to put in a lawn mower and other garden equipment when the children do not need the little house for play.

Children will get a lot of satisfaction out of a rather basic building. For most of their needs, they will not look for much fine detail. You may add trimmings that appeal to adults, but a child might not even notice them. The important items to remember are: a size that the child can enter, but not much bigger than that, a door that the child can close, some spaces to represent windows, and a place the child can call his own. You might get some satisfaction from painting the roof to look like shingles, but a child might be just as happy with a plain, painted roof.

The appearance should be something with which the child can identify. If his home, and all those around, have flat roofs, he might not appreciate a pitched roof. However, a pitched roof is more likely to fit his or her idea of a proper house. From your point of view, it might be better to convince the child that a building with a flat or sloping roof is a real house, if that suits your constructional ideas or method of folding.

Do not overlook safety requirements. Avoid sharp edges and corners. Make sure all woodwork is well-rounded. Be careful of wood which might split or splinter. Some softwood plywood might splinter too easily at the edges. You might have to file the sharpness off edges of hinges and other metal parts. Windows may just be cutouts with rounded edges. Avoid glass. Use soft, transparent plastic, if the child wants more than just a hole. Plain wood turnbuttons are better than more elaborate fasteners. If you arrange a door to fasten inside, make sure you can get at it in an emergency, possibly by reaching through a window or taking the roof off. If you make the building portable, be sure an adult is the only one who operates it. The child should not be able to unfasten anything which would bring the house down on him. Too simple a means of folding might not be a good thing. If you need a wrench on a few nuts and bolts, that puts the action beyond the young experimenting occupant. If you want to provide coat hooks and similar things, it is better to use rounded wooden pegs than nails, which could scratch the child. Be careful that the young users cannot turn the whole house over. If there is no built-in base or floor, you might have to anchor the walls down.

Consider scale. If you make a building to suit a child who is just starting to walk, it can be quite compact, but children grow rapidly. How long do you want the house to last? In only another two years, a child will need much more headroom. He might not mind stooping through a doorway, but he should be able to stand up inside. If you want children up to 10 years of age to use the building, its size must become almost adult. If it is to be a recreational building for young people older than 10, you have to use adult sizes.

BASIC FOLDING PLAYHOUSE

You can make the simplest small playhouse entirely from ½-inch plywood panels, hinging them together so it is portable. The house shown in FIG. 8-1 has a roof that lifts off and ends which are hinged centrally (FIG. 8-2A), so it is possible to fold the parts into a bundle which is under 5 inches thick. The greatest packed length is 48 inches and the greatest packed width is 38 inches.

FIG. 8-1. *You can make a basic folding playhouse from plywood sheets.*

Sizes are arranged so you can cut them economically from 48-inch × 96-inch plywood sheets (FIG. 8-2B). As shown, there are openings for window and door. You could hinge on a plywood door and fit plastic sheet to the window, although for the age child this is intended for, simple openings should be satisfying.

Be careful to square all parts, or they will not fit and fold properly. Make four end pieces (FIG. 8-3A). With these sizes, the roof will slope at about 30 degrees. Make the back and front to match each other and as high as the eaves on the ends (FIG. 8-3B). Cut the door and window openings in the front. Remove sharpness from all edges and round the door and window edges thoroughly.

It should be sufficient to put three 2-inch hinges on each joint. So the parts will fold against each other, the hinges at the centers of the ends come outside (FIG. 8-2C), and those between the ends and the back and front are inside (FIG. 8-2D).

Screws probably will not hold adequately in ½-inch plywood. Although it might be possible to use small nuts and bolts, the neatest way of fixing the

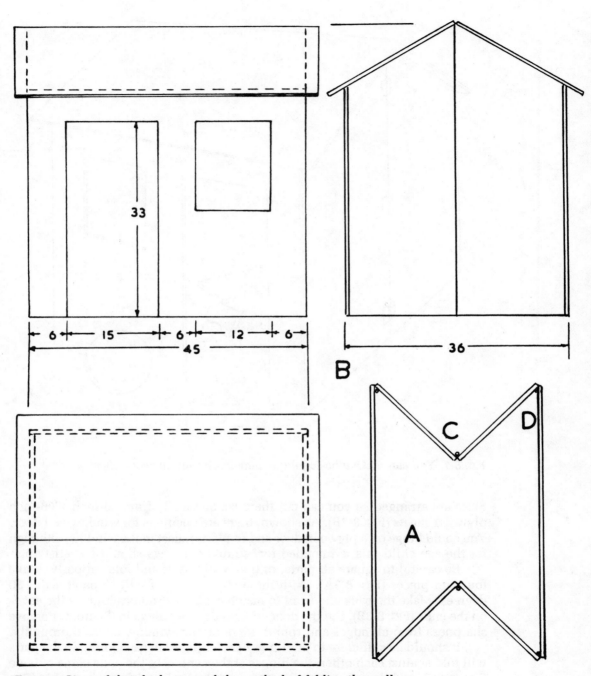

FIG. 8-2. Sizes of the playhouse and the method of folding the walls.

hinges is by rivetting. The rivetting avoids projections which could scratch young hands. You can make suitable rivets with soft-metal nails—copper is particularly suitable. Drill slightly undersize for each nail and drive it through from the other side. Cut off the nail end to leave sufficient length to hammer into the countersunk hole in the hinge (FIG. 8-3C). Support the nail head on an iron block, and work around the projecting end so as to spread it gradually, preferably using a light ball-pane hammer. Try to fill the countersink (FIG. 8-3D). Any excess may be filed off. Adjust the amount the hinge knuckle projects so the ends will open flat and the corners finish close when square to each other.

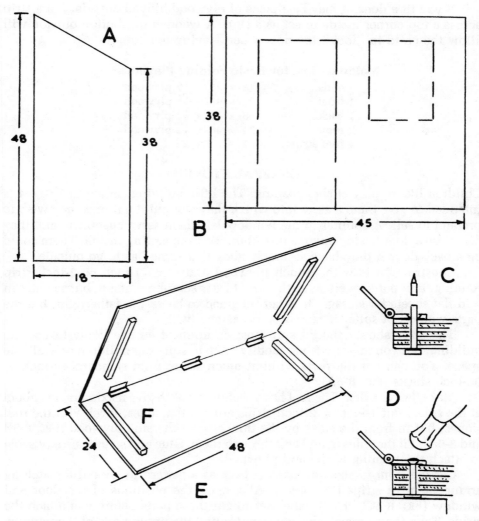

FIG. 8-3. Sizes of the folding playhouse parts and the method of riveting the hinges.

The two roof sections (FIG. 8-3E) will overlap the walls by a small amount. Hinge them together and locate them on the assembled walls so the overhangs are even. Mark the positions of the ends under the roof. Glue and nail strips to fit inside the ends (FIG. 8-3F). Their lower ends should come against the front and back, but the upper ends may be cut back about 2 inches. The roof then will hold the walls in shape. This procedure might be all you need to do, but if the child is able to push the roof, you could fit hooks and eyes outside, under the roof at the ends.

Finish the house in bright colors, with the outside walls a different color from the roof and the inside walls a lighter color. You could edge the openings with a darker color.

If you fit a door, it may be a piece of plywood hinged outside. Put a strip across a top corner inside to act as a stop. A wooden turnbutton outside will allow the child to "lock" the door when leaving the house.

Materials List for Basic Folding Playhouse

4 ends	18- × -48- × -½ plywood
1 front	38- × -45- × -½ plywood
1 back	38- × -45- × -½ plywood
2 roofs	24- × -48- × -½ plywood
4 roof strips	1 × 1 × 21

GENERAL STORE

Children like to play at shopkeeping. The little building shown in FIG. 8-4 is intended to give them a store into which they can put the things they wish to pretend to sell. A counter in the window lets them serve customers and they can close a door to stop anyone unauthorized from getting inside. There could be a back door if they have so much stock that some has to go outside.

Construction is with ½-inch plywood and some 1-inch × 2-inch strip framing. Bolt parts together, so the structure is semi-permanent, but so you can fold flat sheets for storage. It is not intended to be very weathertight, but the parts would not suffer if rained on occasionally.

The sizes shown should suit most children of an age likely to use the building, but you might wish to modify them to suit your children or available space. You can cut the parts without much waste from standard 48-inch × 96-inch sheets (FIG. 8-5A).

Make the back first. The roof has a shallow slope to give maximum headroom at the sides. Put 1-inch × 2-inch strips around the edges, level with the roof slopes, but in from the sides by the thickness of the side plywood (FIG. 8-5B and 8-6A). Fit the sides in and bolt through them. Glue and screws are advisable to attach the framing to the end plywood.

Cut the front to size and use the back as a pattern to lay out a matching arrangement of strips (FIG. 8-6B). Also mark the positions of the door and window (FIG. 8-5C). it will be easier to cut these parts before you attach the strips. Take sharpness off the edges and round the upper parts of the window opening. Cut a door opening in the back, if you wish.

FIG. 8-4. *A child can use this general store to pretend to sell goods.*

The two sides are plain rectangles (FIG. 8-5D). At each corner, drill for three ¼-inch coach bolts. You can make a window opening in one or both sides, but it probably will be better if you leave them solid. In this size building, there should be no need to fasten the roof to the sides, but if you make the store longer from front to back, place a strip matching the slope of the roof along each top edge to take bolts through the roof.

Assemble the parts made so far, so you can test the fit of the roof panels as you make them. Place them against the front, over the strips there. At the back, they may project about 1 inch and at the sides they should project almost to the width of the front (FIG. 8-6C). Hinge the panels together along the ridge, preferably rivetting the hinges in the way described for the folding playhouse (FIG. 8-3C,D). At back and front, drill for two bolts on each slope. These bolts should be enough to keep the roof on and hold the assembly in shape.

You can make the serving counter set into the window opening permanent, but that would interfere with packing the house flat if you disassemble the building. Instead, you could make it with two strips underneath to press over the lower part of the window (FIG. 8-5E). This assembly should be firm enough in use, but you can lift off the counter.

Make a door from a piece of plywood (FIG. 8-6D). Allow easy clearance at the sides, and you can make the bottom up to 1 inch from the ground. A strip inside the top corner will act as a stop and be out of the way (FIG. 8-6E). Rivet

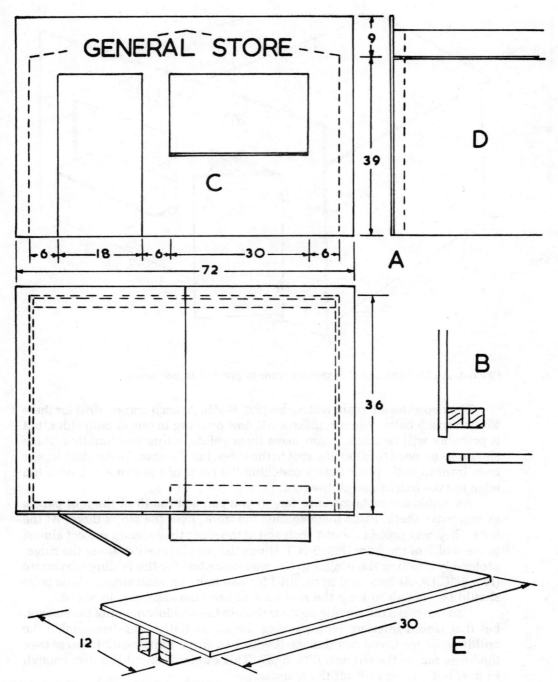

FIG. 8-5. Sizes of parts of the child's general store.

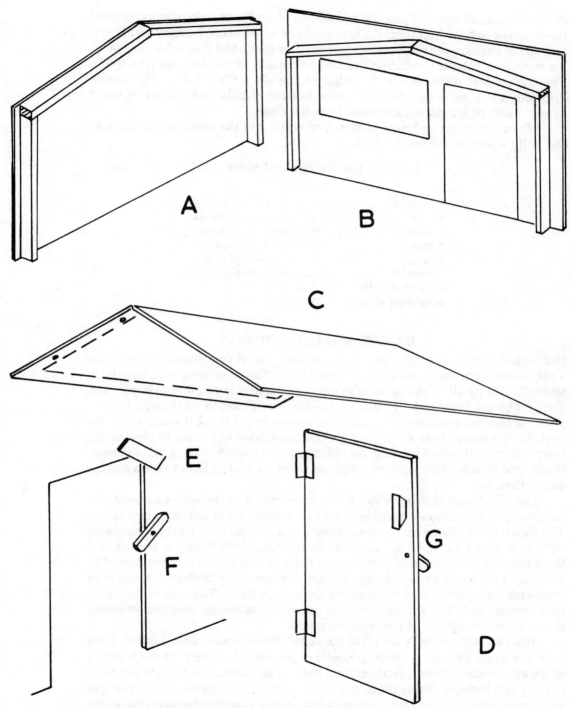

FIG. 8-6. Construction of the parts of the child's general store.

on the hinges, although if you want to screw, place a thin strip of wood behind the doorway to take their points. There could be wooden turnbuttons inside and out, so the shopkeeper can shut out intruders or fasten the door when he leaves his store. The outside turnbutton must be on the side of the doorway (FIG. 8-6F). The inside one must be on the edge of the door (FIG. 8-6G). The inside turnbutton will serve as a handle inside, but you should add a block of wood or some sort of handle on the outside of the door.

When you paint in bright colors, you could put the name of the store or the child's own name on the front.

Materials List for General Store

1 front	48- × -72- × -½ plywood
1 back	45- × -66- × -½ plywood
2 sides	36- × -39- × -½ plywood
2 roofs	36- × -39- × -½ plywood
1 door	18- × -36- × -½ plywood
1 counter	12- × -30- × -½ plywood
2 counter strips	1 × 2 × 30
8 framing strips	1 × 2 × 39

HARDBOARD PLAYHOUSE

Hardboard is a convenient and cheap material, but if you decide to use it for a playhouse, you must understand its limitations. The cheapest grades have little strength. They still might be satisfactory, particularly if you do not expect the demand by the child to be very long. Oil-tempered hardboard is considerably stronger and has a resistance to moisture, so you might think it worth the extra cost. No hardboard is as strong as plywood. As plywood is only ⅛ inch thick, there has to be plenty of framing, including around any window or other cutout. Hardboard does not take paint as simply as wood does—you have to use a special sealer first.

The playhouse shown in FIG. 8-7 is an example of the way you must use hardboard with adequate framing. You can assemble the house in two ways. The framing may be on the inside, so there is a smooth exterior, or it could be outside to give a "Tudor" appearance. As one side of hardboard is smooth and the other side has a textured pattern, you must decide which way you want the hardboard to face. You can change the appearance at any time by turning over front and back panels and exchanging the end sections. The roof will suit either arrangement. You will have to make sure you can change over corner joints, if you want the option of two-way assembly.

You can economically cut all of the panels of the house shown in FIG. 8-8A from the usual 48-inch × 96-inch sheets of hardboard. Nearly all the framing strips are 1-inch × 2-inch sections, laid flat on the hardboard. You do not need to cut joints between frame parts, but if you want to strengthen meeting strips, you could drive in corrugated fasteners (FIG. 8-8B). Use glue between the strips and the hardboard, driving plenty of fine nails from the hardboard into the wood.

FIG. 8-7. You can make this playhouse of framed hardboard.

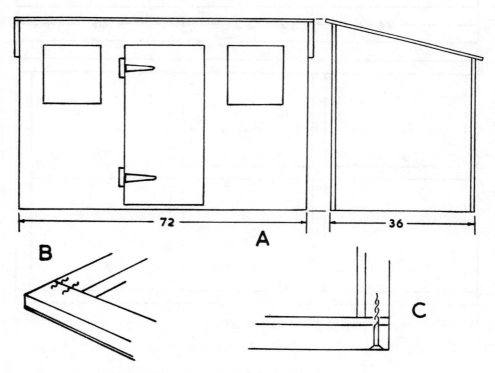

FIG. 8-8. Sizes and construction of the hardboard playhouse.

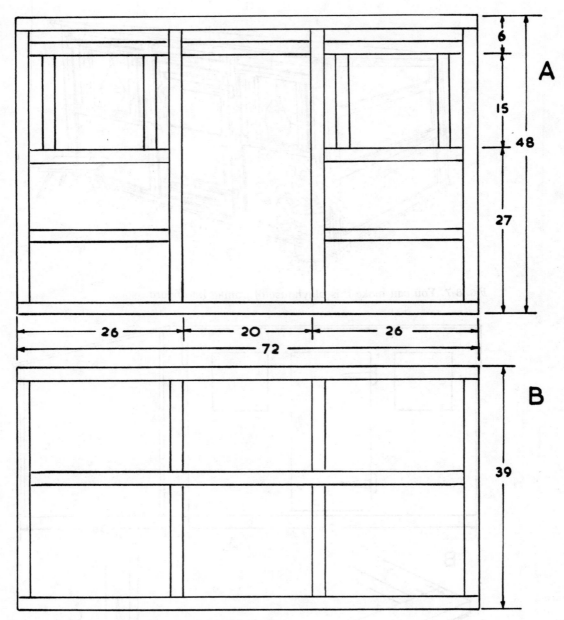

FIG. 8-9. *Front and back of the hardboard playhouse.*

Make the front panel first (FIG. 8-9A). It is shown with a central door and two window openings, but you could alter the arrangement if you wish. Strips are around the edges and every opening. Make the back in a similar way. It is suggested that you do not make any cutouts, and it is 9 inches lower (FIG. 8-9B) than the front.

Make a pair of ends (FIG. 9-10A) to give the width, front to back, that you want. As in the other frames, do not leave too much hardboard unsupported, or an unintentional knock might crack it.

Materials List for Hardboard Playhouse

Front

4 uprights	1 × 2 × 50
4 uprights	1 × 2 × 18
3 rails	1 × 2 × 74
4 rails	1 × 2 × 26

Back

4 uprights	1 × 2 × 41
3 rails	1 × 2 × 74

Ends

1 upright	1 × 2 × 50
1 upright	1 × 2 × 41
4 rails	1 × 2 × 35

Roof

2 pieces	1 × 2 × 78
5 pieces	1 × 2 × 40
1 piece	1 × 1 × 78

Door

2 uprights	1 × 2 × 42
5 rails	1 × 2 × 20

Floor (optional)

4 pieces	1 × 2 × 72
6 pieces	1 × 2 × 36

Covering

⅛ hardboard

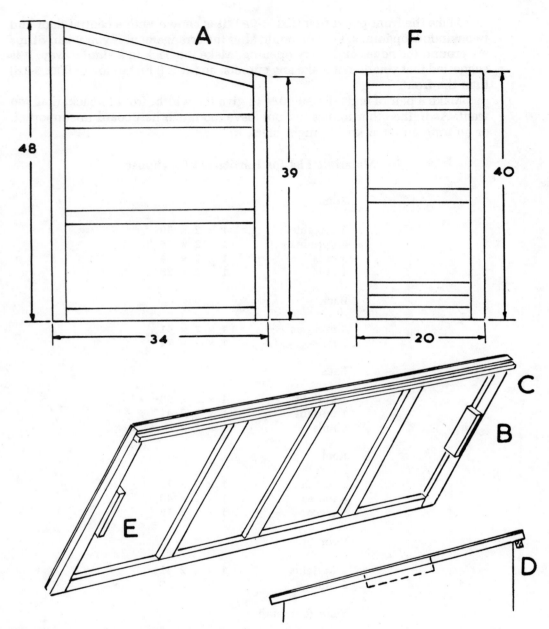

FIG. 8-10. *End, door, and roof of the hardboard playhouse.*

Assemble the parts made so far. Eventually, they will be screwed together with 2½-inch wooden screws at about 9-inch intervals (FIG. 8-8C). For the first assembly, two screws at each corner will be sufficient. You could make a floor to fit inside, if you wish, using hardboard or plywood on framing. The floor would hold the assembly in shape while you make and fit the roof.

Frame the roof like the other parts with the hardboard outside (FIG. 8-10B). Get sizes from the assembled other parts and make it so it overhangs 1½ inches all around. Solid wood then will be bearing on the walls without gaps. Put a 1-inch-square strip along the front (FIG. 8-10C,D) to prevent the roof slipping down the slope, if you do not screw it on. You can place pieces inside the ends (FIG. 8-10E) to locate the roof lengthwise. In the finished house, you might wish to have the roof removable to provide some adult access.

Make the door by framing hardboard (FIG. 8-10F) to fit easily in the doorway. The two crossbars, a short distance in from top and bottom, are to take the hinges, which could be T type about 6 inches long. Arrange a stop in the corner of the doorway. The stop could be a spring or magnetic fastener, but a wooden turnbutton gives a child something to move.

You could glaze the windows with plastic or you could fit them with curtains inside. If the playhouse is permanent or rarely taken apart, you could fit shelves, coat pegs, and other things, such as pictures and cupboards around the walls inside. Children can then treat it as a small home and learn to keep the place tidy.

If you wish to complete the playhouse with the framing exposed on the outside, you can get a traditional effect by having the areas between the framing cream or another light color, while the "timbering" is black or dark brown. Any fairly neutral color will do for the other side, which will then look correct if you ever reverse the walls. The roof might be red-brown or even green, as a simulation of a fullsize roof covering and a contrast to the paint on the walls.

DUAL-PURPOSE PLAYHOUSE

If you want to make a building which children can use for playing and which would be useful for storage for gardening tools as big as lawn mowers, it must be a reasonable size and stronger than anything intended purely as somewhere for younger children to play. It is likely to be permanent, so you must make it weatherproof. The requirements mean its construction will be very similar to other small buildings.

The dual-purpose building shown in FIG. 8-11 has a frontage which you can use as a play area, with a porch. It is large enough for children up to young teenagers, who might use it as a base for games and activities in the vicinity. At the back is a large lift-out door, which gives access to a floor area about 60 inches × 72 inches, to use for storage when children no longer need the building.

As drawn in FIG. 8-12A, the door is almost up to adult height at the front, with a glazed window alongside it. A porch which projects 24 inches shelters the door and window. You can lock the door, so the building then becomes a storage place with wide access at the back. Construction is mostly of 2-inch-square strips covered by shiplap boards.

No floor is shown, but you should place the building on a concrete base. You can put a board floor inside, made with 2-inch-square framing covered with 1-inch boards that project onto the bottom parts of the frames. The frame that crosses under the door is intended to be left in place, but you could cut it away

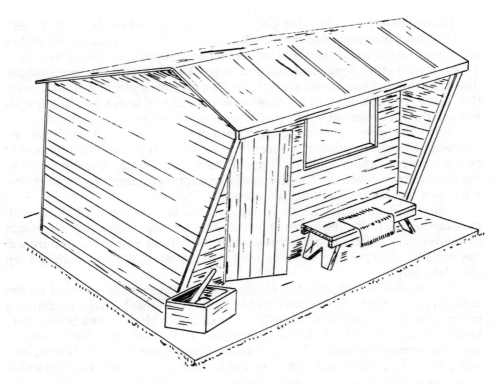

FIG. 8-11. *This roomy playhouse with a sheltered front has a dual use, as a large back door allows you to put gardening equipment inside when the children do not need the house.*

after you have anchored the other parts of the building, if you want a clear floor for wheeling things in.

Start by making a pair of sides (FIG. 8-13A). You can use any of the normal framing joints, except two places need special treatment. At the apex, three pieces meet together. Miter the two sloping pieces together, then halve their meeting ends with the upright (FIG. 8-13B). Place a purlin halfway down the long slope (FIG. 8-13C). This placement requires cutting through the framing member. Put a piece under the cut (which does not go through the covering boards) and square its underside to meet the central upright (FIG. 8-13D).

The front (FIG. 8-14A) fits between the uprights under the apex, so check the height there as you put the frame together. All of the framing is 2-inch-square strips, except for the top, which forms the ridge. The top is 2 inches × 3 inches, and you must bevel to match the slopes of the roof (FIG. 8-14B). If you want the cladding to go close up, bevel that as well.

Carry the uprights to the full height, to give the best support to the horizontal shiplap boarding. Arrange rails between and across them for the doorway and the window. Check squareness by comparing diagonal measurements, as this frame controls the accuracy and symmetry of the building.

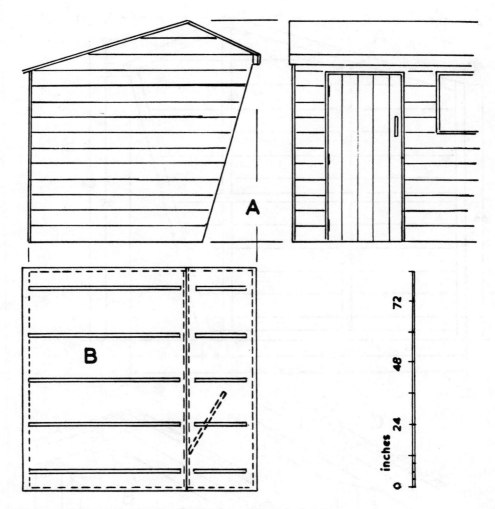

FIG. 8-12. Suggested sizes for the dual-purpose playhouse.

When you erect the building, the joints between the front and the sides will look best if the cladding of the front continues over the uprights on the sides (FIG. 8-14C). Allow enough boarding on each side of the frame to almost cover the adjoining upright.

You can line the building, but assuming it will serve your purpose without lining, edge the doorway and window opening with strips, allowing them to project a little inside and outside (FIG. 8-14D). When you make the back (FIG. 8-15A), its height must match the rear edges of the sides and its width should be the same as the front. Framing is 2-inch-square strips, except the top, which is 2 inches × 3 inches. The doorway is 60 inches wide and 55 inches high. As this doorway takes quite a lot out of the back, it is important that the remaining back is made strongly to prevent distortion of that part of the building. Bevel the top edge to suit the slope of the roof (FIG. 8-15B).

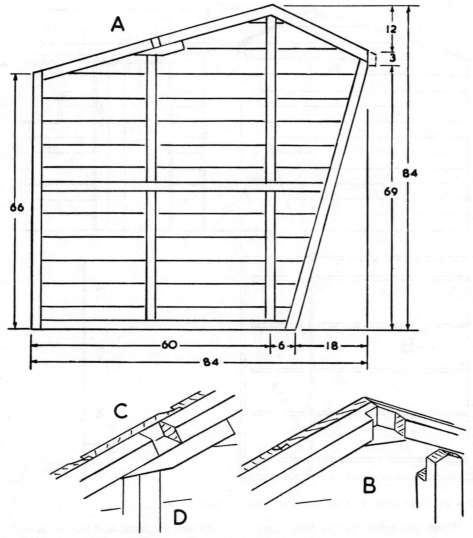

FIG. 8-13. *Details of an end for the dual-purpose playhouse.*

When you assemble the building, the covering boards on the back should overlap the uprights on the sides, to leave a space for a filler piece (FIG. 8-15C). At the top of the doorway, take covering boards to the edge of the door. At the sides, the covering boards on the door must overlap the uprights on the doorway, so cut back the ends of the shiplap boarding on the back to half the thickness of those uprights (FIG. 8-15D).

Make the door frame to fit easily in its opening (FIG. 8-15E). Use 2-inch-square pieces, except for the bottom which is liable to get rougher use. It should be 3 inches deep. Fit diagonals to keep the door in shape. Fit covering boards

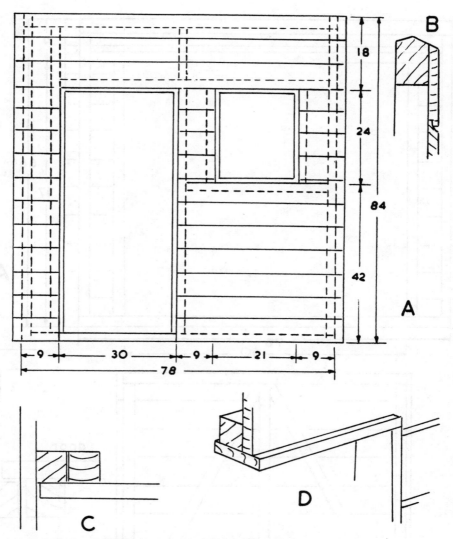

FIG. 8-14. Details of the front of the dual-purpose playhouse.

level at top and bottom, but at the sides extend them to overlap the door uprights, with clearance of the rear of the building is important, cut the door boarding so its lines match the boarding around the doorway.

This extended boarding on the door will prevent the door being pushed inwards. You now have to prevent it from pulling outwards. At the bottom, you can put a board across the width of the door, or make three pegs from 1-inch × 3-inch wood to hook over the bottom member of the back frame (FIG. 8-15F). Taper slightly for easy fitting. Glue and screw to the inside of the bottom of the door. The glue and screws retain the bottom of the door. What you do at the

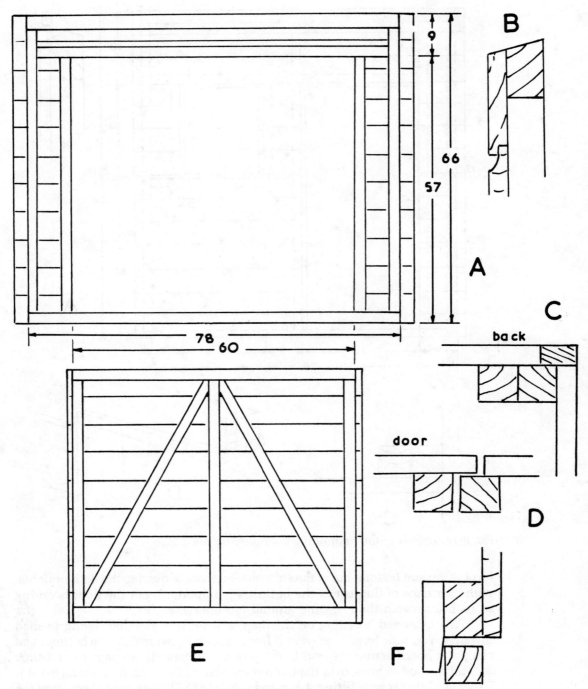

B

A

C

back

door

D

E

F

FIG. 8-15. The back of the dual-purpose playhouse, with a lift-out door.

top depends on the degree of security you desire. The simplest arrangement is a large wooden turnbutton on the back to turn down over the center of the door. You could fit a hasp and staple for a padlock alongside the turnbutton.

You can cover the roof in several ways. You could use corrugated metal or plastic sheets. Plywood would be satisfactory if you covered it with roofing felt or similar material. You can board and cover the roof, and that is the method suggested (FIG. 8-16A).

Assemble the front, back, and sides, using ⅜-inch-coach bolts at about 18-inch intervals. Square the assembly and fasten it down to the base before adding the roof. Fit purlins between the slots in the sides (FIG. 8-16B). Fit the front piece to the ends (FIG. 8-16C). The roof boards extend 2 inches at each side, and the front piece should project the same amount. Cut the boards to meet closely at the ridge (FIG. 8-16D) and project 2 inches at the back. They could be plain boards, tongue-and-groove boards, or shiplap boards laid with the shaped parts underneath. Start at one side and allow for the overhang, then fit further boards tightly.

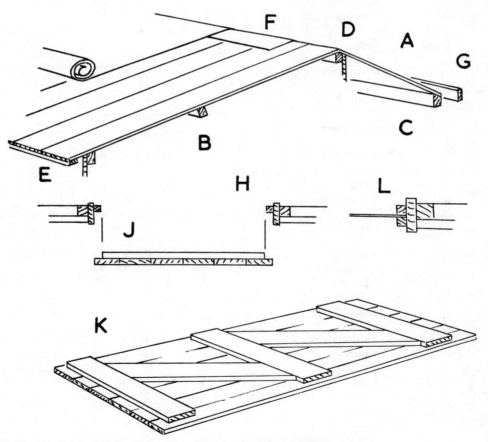

FIG. 8-16. Roof and front door for the dual-purpose playhouse.

Materials List for Dual- Purpose Playhouse

2 uprights	2	×	2	×	86
6 uprights	2	×	2	×	74
2 rails	2	×	2	×	74
4 rails	2	×	2	×	70
2 rails	2	×	2	×	30

Front

4 uprights	2	×	2	×	86
2 rails	2	×	2	×	80
1 rail	1	×	3	×	80
1 rail	2	×	2	×	42
2 window posts	2	×	2	×	30
4 window edges	1	×	4	×	24
8 glazing bars	1	×	1	×	24
2 doorway sides	1	×	4	×	70
1 doorway top	1	×	4	×	32

Back

4 uprights	2	×	2	×	68
2 rails	2	×	2	×	80
1 rail	2	×	3	×	80

Back door

3 uprights	2	×	2	×	58
1 rail	2	×	2	×	62
1 rail	2	×	3	×	62
2 diagonals	2	×	2	×	70

Front door

3 ledges	1	×	6	×	30
2 diagonals	1	×	6	×	40
6 boards	1	×	6	×	70

Roof

1 front strip	2	×	3	×	90
1 purlin	2	×	2	×	86
1 fascia	1	×	5	×	90
1 rear edge	1	×	1	×	90
Covering, boards about	1	×	6		
5 battens	½	×	1	×	65
5 battens	½	×	1	×	26
Cladding, shiplap boards	1	×	6		

Put a strip under the ends of the boards at the back (FIG. 8-16E). Carry roof-covering material right over the ridge from front to back. Put a smaller piece over the ridge area only (FIG. 8-16F), which will give added protection there. Turn the covering material under and nail into the strip at the back. Turn it down at the front and nail there, then cover with a fascia board (FIG. 8-16G). Battens on the slopes (FIG. 8-12B) will prevent the covering material from lifting. It might be sufficient to turn the covering under at the sides, or you can nail battens on the edges there as additional security.

The front door may be ledged and braced, with vertical-plain or tongue-and-groove boards. Put stops around the sides and top of the doorway (FIG. 8-16H), and cut back the ledges on the door to clear them (FIG. 8-16J). Make the door with the bottom ledge high enough to clear the strip across the bottom of the doorway (FIG. 8-16K) and the top ledge a few inches down. Slope the braces up from the hinged side. You can make the door to swing either way. Use plain or T hinges. Either fit a lock or make a turnbutton. Fit a handle on the outside. You might need a catch and handle on the inside. When using the building for storage, you might want a bolt on the inside of the front door, so the only access then is via the rear door.

You could make a framed window to swing open, as described for some other buildings, but a simple fixed window has the glass held between fillets (FIG. 8-16L).

PLAY BARN

A traditional barn has an attractive appearance, and there is an appeal of the olden days to young people, which makes them want to use such a building in the ways they have read about. The scaled-down barn in FIG. 8-17 and 8-18A has the traditional shape with a mansard or gambrel roof. The door is the stable type, so children can close the bottom part and look out above it. A lean-to commonly used as a cart shelter in a fullsize barn, now stores toys up to the size of bicycles. You can use a lean-to at the other side as a porch for sitting out or playing with toys. The barn is large enough for many children to play in it at one time, and it makes a good store for outdoor equipment and the toys which are too large to be taken indoors.

Most of the construction is with shiplap boards on 2-inch-square framing. Make the roof with boards on purlins, then cover with roofing felt or other similar material. The standard design has the door at one end and a window in the other end. You could have doors in both ends. There could be windows in one or both sides, and you might put a door from the barn into the lean-to.

Start with the two ends (FIG. 8-19A), which are the same, except in the closed end the central rail is taken right across (FIG. 8-18B and 8-19B). The boarding goes across below the central rail and the space above becomes the window. The bottom rail across the doorway may stay there in the finished barn, or you can cut it away after you have anchored the building. If you wish to move the barn later, it is better to let the bottom rail in place.

FIG. 8-17. A child-size play barn can provide storage, inside play area, shelter for toys, and a porch to sit under.

Take the shiplap boards to the edges of the doorway. When you erect the barn, the boards from the sides will overlap the end uprights (FIG. 8-19C). For a neat end finish, cut back the boards on the end to the center of each upright (FIG. 8-19D) so you can put a filler piece in each corner (FIG. 8-19E) after you bolt the parts together. Carry the boards to the edges on the roof slopes. Nail the purlins to the edges of the end frames and locate them with cleats, which you can put on now or later (FIG. 8-19F).

If you are to line the building with plywood, handle the ends now. Edge the door and window openings (FIG. 8-19G), going over any lining and extending a short distance inside and out.

The barn sides are simple rectangular frames (FIG. 8-20A). Leave the top edges square. Have the cladding level with the framing at top and bottom, but at the ends, extend the boards enough to overlap the end uprights.

Assemble the ends and sides. Use ¼-inch or ⁵⁄₁₆-inch coach bolts at about 18-inch centers in the corners and cover them with the filler strips. Square the assembly by comparing diagonal measurements, then anchor the building to its base.

Fit the purlins against their cleats to extend about 5 inches at each end, so the bargeboards will be 6 inches from the barn ends (FIG. 8-18C and 8-20B).

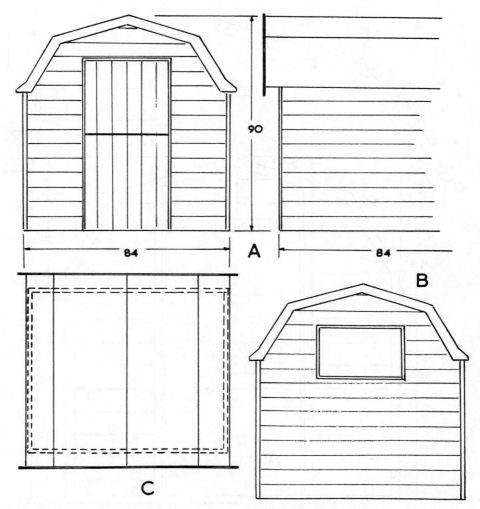

FIG. 8-18. *Sizes and four views of the main building of the play barn.*

Place boards over the purlins with fitted ends where they meet on the ridge and at the angles. You will be covering these joints, but the covering will fit better if you avoid gaps. Boards may be plain or tongue-and-groove. As an alternative, you could use ½-inch or ¾-inch plywood. Put strips of covering material along the joints (FIG. 8-20C). Roll the main covering material from eaves to eaves (FIG. 8-20D), where you should turn it under edge strips and nail it. Turn down at the ends so bargeboards will trap the material. Fit battens at about 18-inch intervals down the slopes on each surface to hold the covering.

Make the bargeboards to stand a little above the surface of the roof and extend below the purlins. Fit the boards to each other (FIG. 8-20E) and take the ends to a few inches below the roof level. Trim the ends parallel with the floor. Triangular additions (FIG. 8-20F) will give the traditional appearance. Gaps

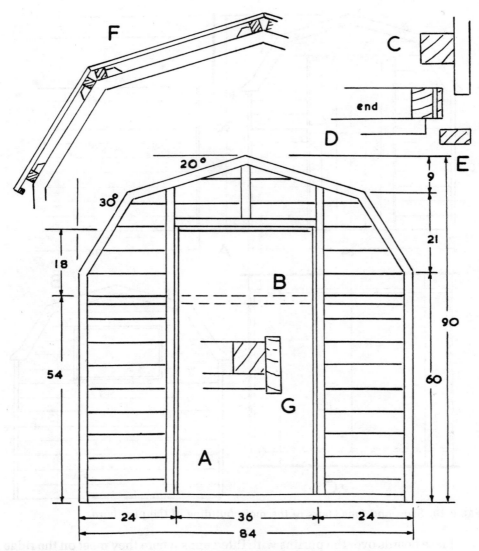

FIG. 8-19. *An end and roof details of the play barn.*

under the roof may be left as ventilation. To close them, you can fit pieces in or continue the cladding to touch the roof.

Make the door in the same way as the two-part door described for the barn in Chapter 7 (FIG. 7-24D) and hinge it in the same way. You can leave the window at the other end open. You could hinge a shutter over it, or you could glaze it, either with glass between fillets, as in the last building, or with a separately framed window that you can hinge open.

The barn is a complete unit. If you are going to add lean-tos, they fit against the cladding, and you can screw or bolt the parts through to framework inside.

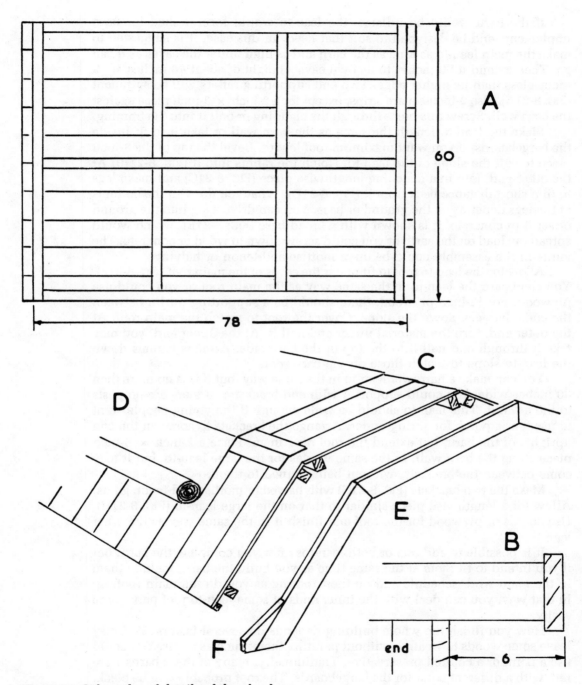

FIG. 8-20. Side and roof details of the play barn.

If the barn would be fullsize, the lean-to would form a shed for farm implements and be designed to suit that need. In this case, it is suggested to make the main lean-to as long as the barn and as high under the eaves as it can go. Then extend it 60 inches to its own eave's height of about 48 inches or 12 inches less than its higher edge. Two end supporting rafters will be sufficient (FIG. 8-21A). Use 2-inch-square strips, except for a 2-inch × 3-inch piece against the barn wall. Screw this piece through the cladding or bolt it into the framing.

Make the lean-to length the same as the barn wall or take it to fit inside the bargeboards, if you want maximum roof length. Bevel the top of the 3-inch piece to suit the slope of the lean-to. Notch the rafters into it (FIG. 8-21B). At the other end, join in a matching lengthwise piece (FIG. 8-21C) and make legs to fit a short distance back from it (FIG. 8-21D). What you do with the bottoms of the legs depends on the ground or base. You can drive a leg into the ground or set it in concrete. It is shown with a broad base (FIG. 8-21E), which would spread the load on the earth or you could screw down to wood or concrete. The joints in the assembly could be open mortise-and-tenon or halving.

Allow for the lean-to roof to fit under the eaves of the main roof (FIG. 8-21F). You can board the lean-to in the same way as the main roof or you could use plywood. For ½-inch plywood, there should be a supporting purlin between the ends, halfway down the slope. Cover the roof to match the main roof. At the outer end, turn the material under and nail it. At the upper end, you may take it through and nail it to the top of the barn side. Arrange battens down the lean-to slope to match those on the barn roof.

You can make a narrower lean-to in the same way, but if it is no more than 30 inches wide, you could support it with end brackets, so there are no posts to the ground. This lean-to should be wide enough if the young people want to use it as a porch for sitting under. Arrange the brackets to come on the end uprights of the barn, but extend the roof over them. Place a 2-inch × 3-inch piece along the barn wall, in the same way as for the large lean-to, but it may come between the brackets, without being joined to them.

Make the two brackets (FIG. 8-21G) with halved or mortise-and-tenon joints. Allow for a lengthwise piece similar to that on the large lean-to (FIG. 8-21H). Use boards or plywood for the roof and finish it in the same way as the other lean-to.

It is possible to add one or both lean-tos after you complete the barn, but if you intend to fit them at the same time as you build the barn, include them in the main work schedule and do their roofing as you do the main roofing. In that way, you can deal with the inner ends of some lean-to roof parts more easily.

How you finish the whole building depends on several factors. You may leave some woods to weather without painting them, such as cedar. You could use a paint or a colored preservative. Traditionally, many of these barns were red, with a different color for the bargeboards. The roof probably will be black. The brighter colors might appeal to young users, but you might wish to choose them to fit in with the surroundings.

Materials List for Play Barn

Door

4 ledges	1	× 6	×	30
2 braces	1	× 6	×	34
2 sides	1	× 3	×	40
2 sides	1	× 3	×	30
6 boards	1	× 6	×	40
6 boards	1	× 6	×	30

Sides

8 uprights	2 × 2		62
6 rails	2 × 2	×	80
4 corner fillers	1 × 2½	×	62

Roof

8 purlins	2 × 2	×	10
8 bargeboards	1 × 5	×	24
16 battens	½ × 1	×	24

Covering

cladding	1-x-6-shiplap boards
roofing	1-x-6 plain or tongue-and-groove boards

Ends

4 uprights	2	× 2	×	90
4 uprights	2	× 2	×	64
2 uprights	2	× 2	×	24
4 rails	2	× 2	×	86
1 rail	2	× 2	×	40
8 rafters	2	× 2	×	30
2 door edges	1	× 4	×	74
1 door edge	1	× 4	×	40
4 window edges	1	× 4	×	40

Large lean-to

1 rail	2	× 3	×	86
1 rail	2	× 2	×	86
2 rafters	2	× 2	×	64
2 legs	2	× 2	×	52

Small lean-to

1 rail	2	× 3	×	86
1 rail	2	× 2	×	86
2 rafters	2	× 2	×	32
2 uprights	2	× 2	×	30
2 braces	2	× 2	×	32

FORT

Young people enjoy playing at pioneer days, with frontier activities and make-believe fights as wagon trains move west, so a playhouse that looks like a fort and stockade obviously would be welcomed. The problem is that such a building would not look right if you made it of squared wood, and there would be no visible place for plywood and manufactured board.

You should make a fort with an authentic look with plenty of natural wood, at least for the external parts. This design means that you need a supply of logs and wood showing the natural bark or outside shapes. If you have access to a good supply of trees to fell or which are already in the form of poles or logs, the making of a fort and stockade playhouse should be easy. Even if none of this material seems possible, you could try a lumberyard or a sawmill for offcuts. When lumber is converted from the log to squared boards, a large number of outside slabs have to be cut away and these pieces might be burned there or sold as firewood or garden decorations. Usually there are more of these offcuts than the sawmill can dispose of, and you might be able to get all you need. Although these pieces are comparatively thin, you can use them to face sawn and

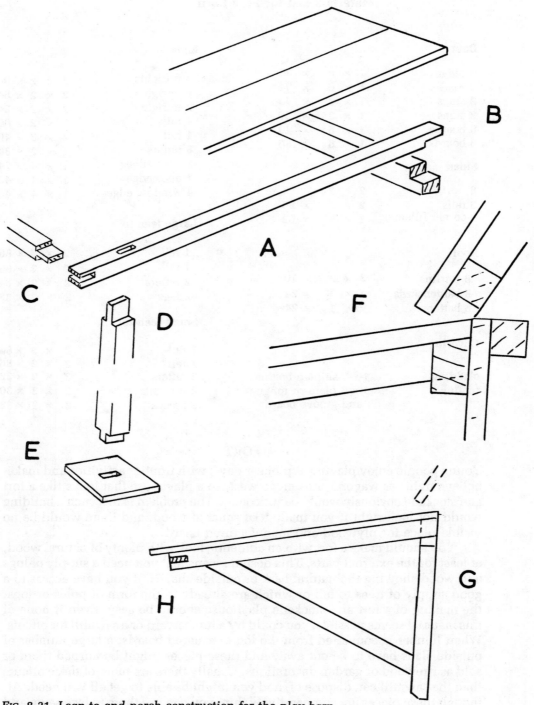

Fig. 8-21. Lean-to and porch construction for the play barn.

squared lumber, so the external appearance gives the effect of a structure built of logs. Even if you have plenty of poles, you will have to cut many of them down the middle, either by splitting (which would have been the original way), or with a circular saw. For close fitting, you can cut some of these pieces parallel (FIG. 8-22A). For a fence, gaps caused by uneven edges might not matter, but for a building, a closer edge fit will be better. This fit is even more important if you are backing up with solid wood or plywood and you do not want it to show through.

It is advisable to peel off bark. With many woods, it comes away easily and the surface exposed still looks natural and rustic. Beside being rough for children to rub against, bark usually hides insects which you do not want, and it could encourage rot. If you want to treat the wood with preservative, you cannot do that effectively if there is bark.

The original stockade had the tops of the logs in a palisade wall sharpened to lethal points (FIG. 8-22B). For a children's fort, they should have less acute angles and you should round or flatten the tops (FIG. 8-22C). A fence will look authentic to the child, but should offer minimum danger if he clambers over.

A stockade wall could have sawn wooden rails, as they would not be very obvious and you could notch them into upright poles driven into the ground (FIG. 8-22D). At a corner, bring the rails in at different levels (FIG. 8-22E).

You could completely lift off some original gates, but for use by children, it would be better to hinge the gates inside. Closure might be by a substantial latch (FIG. 8-22F). Operation from outside might be by a rope through a hole (FIG. 8-22G). Inclusion of this rope is advisable if you want to get in in an emergency and an adult cannot reach over to work the latch. Security against invaders was by a strong bar fitted into sockets (FIG. 8-22H). Knowing your children, you will have to decide if you should provide this.

Within the stockade, children will expect at least one enclosed building, preferably raised (FIG. 8-23). For strength and safety, this building probably is made best of sawn wood and plywood, with a facing to give a log cabin appearance. Some of the posts might also support the fence, but make sure they are secure—a cabin support that sinks or tilts would be difficult to rectify. If the ground does not offer much support unaided, sink a post in concrete, but keep that below the final level, so only soil shows (FIG. 8-24A).

Notch in beams (FIG. 8-24B). If the stockade is 48 inches high, the beams could be 60 inches above the ground. For a building about 72 inches square, you could use 2-inch × 3-inch wood. On this wood, arrange joists of the same section about 18 inches apart and put ½-inch plywood on them (FIG. 8-24C).

Make the roof in a rather similar way, with beams across two pairs of poles and rafters and plywood over them (FIG. 8-24D). The slope need only be slight —12 inches in 72 inches would be satisfactory.

Arrange walls at a height that will stop children from falling out—30 inches probably will do. Use plywood again on rails (FIG. 8-24E). At the entrance, put a post to the full height to support the wall and provide something on which to grab. Keep the entrance fairly narrow for safety.

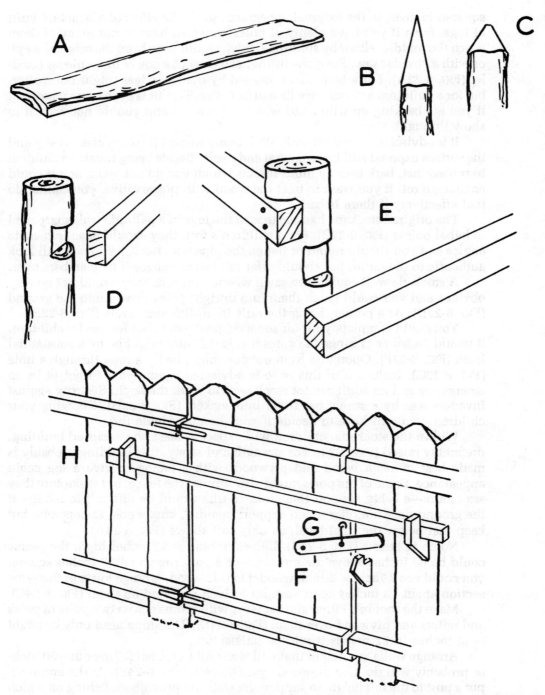

FIG. 8-22. The method of making a stockade fence for a child's fort. Avoid sharpness (B). Make blunt ends (C).

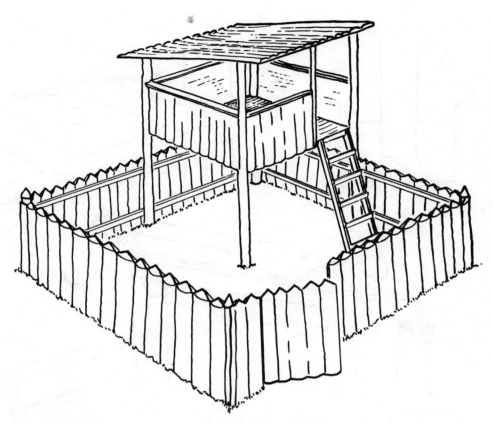

FIG. 8-23. A possible layout for a child's fort and stockade.

Boxing in with plywood will provide safety, but you have to give the outside the correct appearance by covering with slabs of wood for the natural exterior (FIG. 8-24F). Cover tops for a safe edge to lean against. With squared wood, you can give it a more primitive appearance by scooping hollows out of the edges (FIG. 8-24G).

The access ladder could be upright, but it is easier to climb if it slopes. Attach securely at the top and let the bottom into the ground. Notch in rungs at about 8-inch intervals (FIG. 8-24H), with the same stepping distance at both ends (FIG. 8-24J), so an unexpected step height will not cause the child to trip. You can use waney-edge wood for the ladder sides, but if you cannot find suitable pieces, you may scoop the edges to simulate a more natural appearance.

There could be a gate opening inwards at the entrance. A spring closure would aid safety, even if the original defenders of the fort would not have known such things.

You can add several things. A raised platform inside the stockade fence would serve as a step for looking over and as a seat. Do not make it so high that a child could overbalance outwards. Holes in the fence would allow smaller

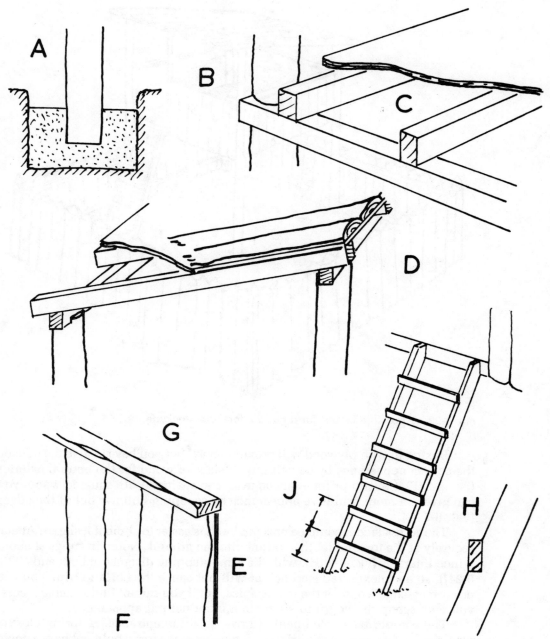

FIG. 8-24. Constructional details for the fort.

children to look out. You can cut such gun slots in the cabin sides. If you build in benches in the cabin, be careful that a child standing on one could not topple over the side. It might be safer to arrange benches at ground level under the cabin.

If the fort is to stand for long, treat all the wood with preservative. Decay in the posts is most likely at ground level. Use plenty of preservative down to the concrete, before covering with soil. Some preservatives take a long time to dry, so you should keep children away during drying, and any renewal of preservative is best done at a time when the fort is not needed, at least for a few weeks.

GLOSSARY

The making of wooden buildings is only part of the much wider craft of woodworking. The selection of words that follows include some that are particularly appropriate to the subject of this book, and might be helpful to readers unfamiliar with the language of this branch of the craft.

ark—A building for small animals.
aviary—A building to hold birds.

bargeboards—Covering boards at the end of a roof, usually a gable.
battens—Strips of wood of light section, used on a roof to hold down the covering.
bay—A space or section in a building.
brace—A diagonal strut used to triangulate an assembly and prevent it from distorting.

chipboard—A board made by bonding wood chips with a synthetic resin. Also called *particleboard*.
cladding—The covering of boards that forms the outside of a wooden wall.
cleat—A link between other parts. A strip of wood across other pieces. A support for a purlin on a rafter.
coach bolt—A bolt for use with a nut, having a shallow, round head and a square neck to grip the wood. Also called a *carriage bolt*.
coach screw—A large wooden screw with a square neck for use with a wrench. Also called a *lag screw*.
corrugated fastener—A sheet-steel fastener with an undulating cross-section, sharpened at one side and used like a nail to drive into adjoining pieces of wood to hold them together.
corrugated sheet—Roofing material, which might be steel, other metal, or plastic. The corrugations in the length provide stiffness.
cut nail—A nail made from sheet steel instead of the more usual wire.

dead pin—A wedge or dowel.

drip groove—A groove cut along the underside of a sill or other projecting wood to prevent water running back along the surface.

eaves—The angle between a roof and a wall. The overhang of a roof over a wall. Never spelled without the s, even if there is only one.

exterior-grade plywood—Plywood in which the glue used is waterproof.

eye screw—A wooden screw with a ring or eye as head. Also called a *screw eye*.

fascia—1. A long, flat, wooden surface, such as at the front of a lean-to roof. 2. The casing of a door.

gable—The end of a roof, usually one with a ridge.

galvanizing—A method of coating steel with zinc as a protection against rust. Used on corrugated steel roof sheets.

gambrel roof—A roof with two slopes on each side. Also called a *mansard roof*.

gazebo—A structure intended to be decorative and from which you can obtain a view.

hardboard—A thin board made from compressed shredded wood.

hardwood—Wood mainly from broad-leafed trees which shed their leaves in the winter.

hip roof—The end of a ridge roof that slopes inward instead of having a vertical gable.

joist—A supporting beam, as in a floor.

lag screw—Alternative name for a *coach screw*.

lean-to—A building with a roof having a single slope, which could be against another building or stand free.

ledge (ledged)—A piece across a door made of vertical boards, used with braces to keep the door in shape.

mansard roof—A roof with two slopes on each side. Also called a *gambrel roof*.

nominal—When applied to lumber, of or referring to the sawn size and the planed size is smaller.

palisade—A fence of upright boards.

particleboard—A board made by bonding wood chips with a synthetic resin. Also called *chipboard*.

pergola—An arbor or covered walk with the wooden structure covered with growing plants.

plywood—Manufactured board made by gluing thin pieces of wood with the grain of alternate pieces arranged at right angles.

pole construction—A barn or other building made with poles as the main structural parts.

pop hole—The entrance for poultry into their house.

purlin—A lengthwise support for roof covering, usually supported on rafters.

rabbet (rebate)—Recess in the edge of wood, as in a picture frame.

rafter—A support for a roof.

rail—A horizontal structural member.

ridge roof—A roof with an apex, like an inverted V.

roofing felt—A flexible roof covering material to lay over boards, made of felt impregnated with tar or other waterproof substance.

roof truss—A braced framework with rafters, for supporting a roof.

screw eye—A wood screw with a ring or eye head. Also called an *eye screw*.

shiplap boards—Cladding boards, to be laid horizontally, with the upper edge of each one fitting into a rabbet in the one above.

sill (cill)—A projecting horizontal board, such as at the bottom of a window, to shed water away from the wall below.

softwood—Wood from needle-leaf trees.

stable door—A door in two parts, so you can open the top part while the lower part remains closed.

stainless steel—Steel with other metals alloyed to it to give it a resistance to corrosion.

staple—A nail in the form of a U with double points.

stove bolt—A bolt threaded to the head, which has a screwdriver slot.

stressed skin—An assembly in which the skin plays a major part in providing strength.

sway bracing—Diagonal pieces arranged in an assembly where the parts cross squarely, to triangulate it and provide resistance to distortion, particularly in strong winds.

tack—1. A small nail tapered for most of its length. 2. Harness and other equipment used with a horse.

tempered hardboard—Hardboard treated with oil, which strengthens it and gives it a resistance to water.

tie—A member in a structure intended to hold parts together and resist stretching loads, as in a roof truss, where it prevents the rafters spreading.

tongue-and-groove boards—Boards prepared so a tongue on the edge of one piece fits the groove on the edge of the next piece.

triangulation—Placing a member diagonally across a four-sided figure to divide it into triangles, so it keeps its shape.

truss—Supporting structural framework. In a building, rafters might be supported by a truss.

waney edge—The shape of the outside of the tree on a board that has not had its edge squared.

weatherboarding—Cladding boards to be laid horizontally, in the same way as shiplap boards, but tapered in the width so the thin edge of a lower board goes under the thicker lower edge of the one above it.

wind bracing—A type of sway bracing, but arranged in the roof, diagonally between trusses or purlins, to protect the roof from loads resulting from strong winds.

INDEX

INDEX

INDEX